Praise for Mysteries of the First Instant

"Understanding space, time, the forces of nature, and elementary particles finally becomes easy. Written like a novel, *Mysteries of the First Instant* brings to a sublime level Friedmann's passion for pinpointing scripture in science, and science in scripture. It is more than a novel or a treatise — it's an emanation."
— David W. Menefee, author

"Congratulations! As far as I know, this is the first theologically-oriented analysis of the Genesis Creation narrative that I have seen that attempts to view it in the light of the findings of modern physics."
— Dr. David C. Bossard, author

"In an accessible and well-illustrated and documented manner of presentation, this book poses answers that many have never considered. The journey of reading this superb book becomes a thought-provoking experience unlike any other book. Very highly recommended."
— Grady Harp, Amazon HALL OF FAME; TOP 100 REVIEWER

"This book is a wonderful and very educational read. Using the simplest examples, the author leads us to the understanding of the main postulates."

"I loved the way this book engaged me as a reader and a learner to discover new ideas about the beginning of the universe."

"This book was the best combination of Scientific Studies and Religious theology. I appreciate the author's focus to give information in a simple manner so the reader can draw their own conclusions by the end of this book."
— Amazon readers

Other Books by Daniel Friedmann
Inspired Studies Series

The Genesis One Code demonstrates an alignment between the dates of key events in the development of the universe and the appearance of life on Earth as described in Chapters 1 and 2 of Genesis and as derived from scientific theory and observation.

The Broken Gift follows and extends the scope of *The Genesis One Code* to include the appearance and early history of humans as described in Chapters 2 to 11 of Genesis, comparing the biblical chronology with dates derived from scientific theory and observation.

Roadmap to the End of Days explains the biblical timeline of human history and for the End of Days, placing recent history in context and providing a glimpse of what is coming next.

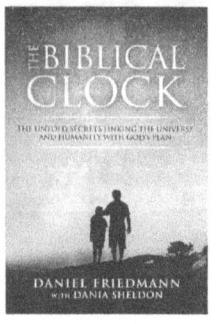

Co-authored with Dania Sheldon, *The Biblical Clock* is a narrative describing Friedmann's quest for answers that produced the prior three books. It is a standalone volume and in relating the story of discovery includes the essential materials in the earlier trilogy.

Other Books by Daniel Friedmann
Origins Series

Daniel Friedmann's prior books explain key events in the history of the universe, when they happened, and what is likely to happen next here on Earth.

But what about how it all started?

<u>Mysteries of the First Instant</u>: Illuminating What Science Hasn't Answered about the Inception of Our Universe

Through scientific discoveries, we know in exquisite detail about the universe's development after its initial moment of inception, yet we've remained in the dark about how it began and why the laws of physics have their specific characteristics.

Until now.

Mysteries of the First Instant reveals that the esoteric dimension of biblical texts teaches the deepest cosmology and contains ultimate insights into the fundamental workings of the universe.

<u>Mysteries of the Origin of Animals</u>: Illuminating What Science Hasn't Answered about the Inception and Development of Animal Life

Scientists have struggled to explain key events in the appearance and early development of life on Earth. In *Mysteries of the Origin of Animals*, Daniel Friedmann shows how faith and science together help us understand how life arose, how animals appeared and developed, and how they came to have sophisticated body plans, blood, and specific behavioral characteristics, such as foraging or hunting.

Read Chapter 1 of *Mysteries of the Origin of Animals* at the end of this book.

To be notified of the book's publication follow Daniel Friedmann here https://www.amazon.com/Daniel-Friedmann/e/B005LKD1Q4/

Or, please go to https://www.danielfriedmannbooks.ca/contact/

ORIGINS SERIES

Mysteries of the First Instant

Illuminating What Science Hasn't Answered about the Inception of Our Universe

DANIEL FRIEDMANN
with
DANIA SHELDON

Inspired Books

Copyright © 2021

Daniel Friedmann
All rights reserved.

No part of this book may be used or reproduced in any manner whatsoever without the written permission of the authors.

Website: www.danielfriedmannbooks.ca
E-mail: friedmann.edaniel@gmail.com

Website: www.daniasheldon.com
E-mail: dania.sheldon@icloud.com

Cover Design by James
GoOnWrite.com

ISBN: 9781689226691

TABLE OF CONTENTS

A NOTE FROM THE AUTHOR v
THE ORGANIZATION OF THIS BOOK vii
ACKNOWLEDGMENTS viii

PART 1 1

CHAPTER 1 BUILDING BLOCKS 3
CHAPTER 2 EXPANDING ON EINSTEIN 17
CHAPTER 3 HER DARK MATERIALS 25
CHAPTER 4 STAY TUNED 35
CHAPTER 5 GALLOPING INFLATION 45
CHAPTER 6 SOMETHING TOTALLY USELESS 53
CHAPTER 7 MASTER OF ALL THE FORCES 67
CHAPTER 8 BALANCING THE SCALES 81
CHAPTER 9 BROKEN VESSELS 95
CHAPTER 10 IT'S ABOUT TIME 111
CHAPTER 11 FORCES TO BE RECKONED WITH 120
CHAPTER 12 THE SIGNATURE OF GOD 129
CHAPTER 13 SUMMING UP 139
    COSMIC HISTORY 139
    THE FIRST INSTANT 139

PART 2 157

CHAPTER 14 BEFORE THE BIG BANG 159
    THE SCIENCE ANSWER 159
    THE BIBLICAL ANSWER 160
        The Eternal—God *161*
        In the Beginning *162*
CHAPTER 15 TIME AND RELATIVITY 169
    THE SCIENCE ANSWER 169
    THE BIBLICAL ANSWER 171
        Asman and Seder Asman *171*
        The Arrow of Time *173*
        The Role of Time *175*
        Timeless Light *176*

CHAPTER 16 WHEN MICROSCOPIC MEETS MACROSCOPIC 178
THE SCIENCE ANSWER 178
THE BIBLICAL ANSWER 180
The Worlds of Tohu and Tikkun 181
How Do These Worlds Interact? 182
String Theory? 184
The Big Bang and Black Holes 185
CHAPTER 17 ELEMENTARY PARTICLES 186
THE SCIENCE ANSWER 186
THE BIBLICAL ANSWER 187
God's Speech 187
Organization of the Letters 188
Information in the Letters 188
CHAPTER 18 A THEORY OF EVERYTHING? 191
UNITY IN NATURE 193

THANK YOU FOR READING THIS BOOK! 195
HAVE A QUESTION FOR DANIEL? 195
ABOUT THE AUTHORS 196
DOWNLOAD BOOKS 1, 2, 3, AND 4 197
READ AN EXCERPT FROM THE NEXT BOOK IN THE ORIGINS SERIES 198

APPENDIX A THE STANDARD MODEL 199
APPENDIX B THE LIGHT METAPHOR 201
The Language Applied to God 202
The Light Metaphor Extended 203
APPENDIX C THE HEBREW LETTERS AND THEIR NUMERICAL VALUES 204

GLOSSARY 205

CHAPTER RESOURCES 228

IMAGE CREDITS 236

EXCERPT MYSTERIES OF THE ORIGIN OF ANIMALS— CHAPTER 1 242

ENDNOTES 249

FIGURES

Figure 1.1	Illustration from Godtfred Christiansen's patent application	6
Figure 1.2	The constituents of matter	9
Figure 2.1	Einstein's blackboard	19
Figure 2.2	The composition of deuterium and helium	22
Figure 2.3	The cosmic timeline	24
Figure 3.1	The Milky Way, enclosed in a galactic halo of dark matter	27
Figure 3.2	Vera Rubin standing by the Lowell Observatory	28
Figure 3.3	Picture of a section of the universe	30
Figure 3.4	Constituents of the universe	33
Figure 5.1	Solving the flatness problem	49
Figure 6.1	Large Hadron Collider at CERN	56
Figure 6.2	The Standard Model of particle physics	61
Figure 6.3	The laws of physics	63
Figure 7.1	Title page of the first printed edition of the *Zohar*	69
Figure 7.2	Rabbi Chaim Vital	72
Figure 7.3	The first three verses of Genesis	75
Figure 7.4	The Hebrew alphabet and the numerical values of the letters	77
Figure 8.1	The last Hubble repair mission	82
Figure 8.2	The biblical and scientific chronologies	85
Figure 8.3	Rashi	88
Figure 8.4	A page of the Talmud	90
Figure 8.5	Chronology of the universe	92
Figure 9.1	Manuscript of *Etz Hayim*, c. 1770	97
Figure 9.2	The first verse of Genesis, in English, Hebrew script, and transliterated Hebrew	99
Figure 9.3	The second verse of Genesis	101
Figure 9.4	The third verse of Genesis	103
Figure 9.5	The first instant, modified for the void being nonphysical	105

FIGURE 9.6 THE BIBLICAL AND SCIENTIFIC CHRONOLOGIES 106
FIGURE 10.1 THE WAVELENGTH OF LIGHT EXPANDS AS THE UNIVERSE EXPANDS. 113
FIGURE 10.2 THE PLANE OF THE ECLIPTIC 115
FIGURE 10.3 THE ANISOTROPY IN THE COSMIC MICROWAVE BACKGROUND 116
FIGURE 11.1 *SEFER YETZIRAH*, 1562: COVER AND INSIDE PAGE 121
FIGURE 11.2 CHARACTERISTICS OF THE WORLDS OF TOHU AND TIKKUN 124
FIGURE 11.3 THE LETTERS OF GOD'S NAME AND THE FORCES OF NATURE 127
FIGURE 11.4 NEWTON'S LAW OF GRAVITATION AND COULOMB'S LAW OF ELECTROSTATICS 128
FIGURE 12.1 THE HEBREW LETTERS AND THE ELEMENTARY PARTICLES 131
FIGURE 12.2 THE LETTER ALEF, MADE OF THREE LETTERS 133
FIGURE 12.3 PARTICLE AND ANTIPARTICLE 138
FIGURE 13.1 BIBLICAL ELUCIDATION OF THE FIRST INSTANT 151
FIGURE 15.1 TIME MANIFESTS FROM THE CHARACTERISTICS OF THE SIX DAYS OF CREATION TO THE DETAILS OF A FRACTION OF A SECOND 175
FIGURE 17.1 PREDICTED PARTICLE MASSES CORRESPONDING TO HEBREW LETTERS 190

A Note from the Author

When I first read the Genesis account of the origins of the universe, I saw that it provided a chronology of our world coming into being. Yet it didn't seem to address the beginning, other than with "*In the beginning of God's creation of the heavens and the earth.*"[1] But how did He create? Starting with what?

When I came to study engineering physics, it became apparent that science also had a good chronology for the development of the universe. Yet it too didn't seem to address the beginning, other than with the "Big Bang". But what exactly was the bang? What came before it? Did it happen right at the beginning? Later, I learned that the Big Bang theory delineates cosmic evolution from a split second after whatever happened to bring the universe into existence, but it says nothing at all about time zero itself.[2] Further study revealed that physics is unclear about the origins of everything included in the beginning: time, space, elementary particles, and the forces of nature.

So, although popular belief now has it that science explains how the universe came from nothing, this turns out not to be the case. The "nothing" is actually at least space and gravity, and often a state—called the quantum vacuum—with energy and myriad virtual particles coming in and out of existence. The quantum vacuum exists in time and space, and the energy and particles obey the laws of physics. Furthermore, the Big Bang theory does not comment about their origin or explain why initially, the universe was immensely dense and extremely hot.

Normally, further scientific research clarifies what is not known, but when it comes to the beginning of the universe, further research seems to be deepening the mystery. Indeed, we have discovered that the twenty-six or more measured parameters required to drive the Big Bang theory, such as the properties of elementary particles and the strengths of the forces of nature, are very finely and precisely tuned to allow the universe and life to exist.

So, I went back to Genesis and found that most commentaries agree that "in the beginning" does mean the creation of all the building blocks required for the rest of the Genesis account: time, space, elementary particles, and the forces of nature. But where, I wondered, are the details about how all this came about? The mystical interpretation of the Bible includes a detailed cosmology about the beginning, contained in a series of texts starting with the *Book of Creation*—a couple of thousand words long and attributed to Abraham (circa 1740 BCE)—and continuing to hundreds of pages of modern texts elucidating earlier texts.

Comparing scientific theory with biblical cosmology seemed futile. The language is so different, the approach so dissimilar, and the concept of how things come about—supernaturally versus naturally—diametrically opposite. Nonetheless, the study of both revealed a parallel story. Science and the Bible tell us what happened after the first instant and agree that everything can be understood as occurring naturally; that is, there was a chain of events based on cause and effect, operating under the laws of nature.[3] However, the Bible also tells us what happened at the first instant and why the universe has its particular characteristics. As you read this book, you'll see what science has discovered about how the universe came to be, and what the Bible elucidates.

To delve deeply into the subject's mysteries, I have relied mostly on biblical sources shared by the Abrahamic religions, but I elucidate them with Jewish sources because my religious education is based in Judaism. If you have a different religious background, or none, please continue to read. You will find that the various sources pertaining to origins have more in common than you expect.

Every attempt has been made to ensure that no background in cosmology or the Bible is required to understand this book.

The Organization of This Book

Part 1 is a narrative covering three main topics: 1. the scientific approach and the biblical approach to understanding the universe; 2. a review of what science has learned about how the universe came to be, from time zero onward, and how nature behaves at the most fundamental level; and 3. what the Bible says about the universe's creation, particularly the beginning, including what it says we can and can't learn via the scientific method. A final chapter compares the biblical wisdom with scientific discovery, summarizing everything presented in Part 1.

Part 2 devotes a chapter to each of the things we don't yet understand. By elaborating upon the knowledge gained in Part 1, we will go deeper into each of these mysteries. These chapters are largely standalone, so if you're not interested in every topic, you can read just the chapters covering the mysteries that draw your interest.

Finally, the text is backed up by an extensive glossary, numbered references for key statements, and sources for the narrative sections of the chapters in Part 1.

Acknowledgments

Many people were instrumental in helping me with this work. My teachers of many years, Rabbi Avraham Feigelstock and Rabbi Shmuel Yeshayahu, introduced me to key concepts and helped me locate biblical references.

Gabriel Hirsch, Paul Lim, David W. Menefee, David C. Bossard, Eduard Fischer, Dr. Joseph N. Trachtman, and Jeff Cox provided valuable editorial comments, corrections, and feedback.

I would also like to express my sincerest gratitude for the generous help and advice of my wife, Marilyn, who also patiently formatted the manuscript.

Part 1

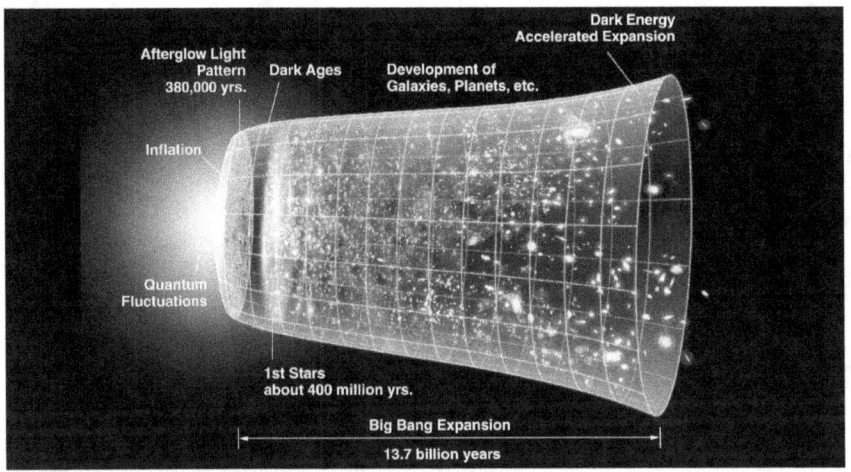

Timeline of the expansion of space

Chapter 1

Building Blocks

Billund, Denmark, February 1960

By the time Godtfred and Kjerk reached the site, flames were shooting a hundred feet into the night sky and it was clear that nothing but the foundations would be left. Firefighters sprayed the roaring blaze to keep it contained, knowing that it would die out only when its fuel was consumed. As the factory and its adjacent warehouse contained thousands of pounds of tinder-dry wood, they were resigned to being there through most of the morning that had not yet begun to dawn.

Godtfred had been four when the first fire had leveled his father's business in the little town of Billund, Denmark. Actually, it had been Godtfred and his older brother, Karl Georg, who had accidentally started the fire when they lit the glue heater; some wood shavings ignited, and the building was razed.

In 1924, it had been an incredibly daunting task for Ole Kirk Christiansen to recover from such a devastating blow. But he was responsible for looking after his wife, Kirstine, and their young children, so Ole had applied his resourceful mind and impressive work ethic to recover from the setback. Godtfred was working in the business by age twelve, helping to make and distribute the high-quality wooden toys their small business was becoming known for. Already, he was showing the same creativity and drive as his father.

Less than two years later, while the business continued to grow, Kirstine died giving birth to a fourth child. Ole and the children were devastated, but they had to keep going, so father and eldest son continued with plans for expanding their toy lines and their sales reach. Now that Lego was becoming increasingly well known in their small country, the next step was to look overseas.

And then, in 1942, fire had struck again. By this time, twenty-two-year-old Godtfred had returned from studying in technical college and was taking on more duties at the company. The two men also felt responsible for the well-being of the employees and their families, so once more, they rolled up their sleeves and worked steadily to rebuild.

By the 1940s, plastic was increasingly used in the manufacturing sector, including in the making of toys. Ole and Godtfred took the plunge in 1947 and became the first to bring a plastic molding machine to Denmark. The making of plastic Lego products had begun. But it took a little while for them to take the form that is now globally known and has accompanied several generations through childhood.

First, Lego began to produce simple plastic toys in addition to their wooden toys. By this time, it was 1953, and Lego had continued to expand. The following year, returning from a toy exhibition in Britain, Godtfred had what would prove to be a pivotal conversation with another passenger in the toy industry. That gentleman opined that although an abundance of toys were being produced in America and Europe, they lacked "a system." What he meant was that companies were producing complete, unchangeable toys, which weren't engaging kids' imaginations. He wanted a system—something that would make children's creativity part of their toys.

This conversation made such an impression upon Godtfred that he began considering how he could bring a system concept to Lego. If he wasn't going to produce a finished toy, what would he produce? When purchasing the plastic molding machine, he had been shown how it could produce a little brick. Perhaps it was the combination of seeing this demonstration plus remembering his father having to rebuild their factory twice. In any case, Godtfred decided that bricks would be the foundation of Lego's new toy system. A relatively small number of brick types would be sufficient to keep kids building and playing indefinitely.

He tried the bricks on his kids. They were an instant hit, and soon the kids were building quite complex structures. But when they tried to move or modify these structures, everything came tumbling down, much to their frustration. Godtfred then realized that having elementary building blocks wasn't enough; there had to be a way to

Chapter 1 – Building Blocks

bind them together semi-permanently so the structure wouldn't fall apart but could still be modified. It wasn't long before he came up with a stud-and-tube coupling system that locked them together securely but not too tightly for children's fingers to pry apart.

Now, everything built with Lego blocks was thoroughly stable and easily expanded and modified. Within months, Godtfred had developed what came to be known as the Lego System in Play. As the company's website explains, "The LEGO System means that: all elements fit together, can be used in multiple ways, can be built together. This means that bricks bought years ago will fit perfectly with bricks bought in the future… [So] all bricks—from yesterday, today and tomorrow—fit together."[1] After that, it seemed the sky was the limit for Lego.

And now, thought Godtfred, surveying the billowing smoke, a huge chunk of Lego's inventory and equipment was heading skyward. He took a deep breath, then exhaled slowly and put a hand on the shoulder of his thirteen-year-old son, Kjerk. "Well, at least your grandfather didn't have to see the company go through a third fire."

"What do we do now, Dad?"

"The same thing one of the kids who enjoy our toys would do if someone came along and smashed their creation into bricks: we gather up the pieces and rebuild. And I think it's time we stopped making wooden toys."

≈≈≈

I was relating this story to my teenage nephew Seb as we strolled through Legoland in Carlsbad, California. After retiring from full-time work at an aerospace company, I had more time to spend with him and other family members, and although we both still had hectic schedules, he and I carved out time for the occasional vacation together. Last year, it had been a week-long kayaking trip in southwestern British Columbia. This year, we had headed for the far south coast of the Golden State to do some surfing.

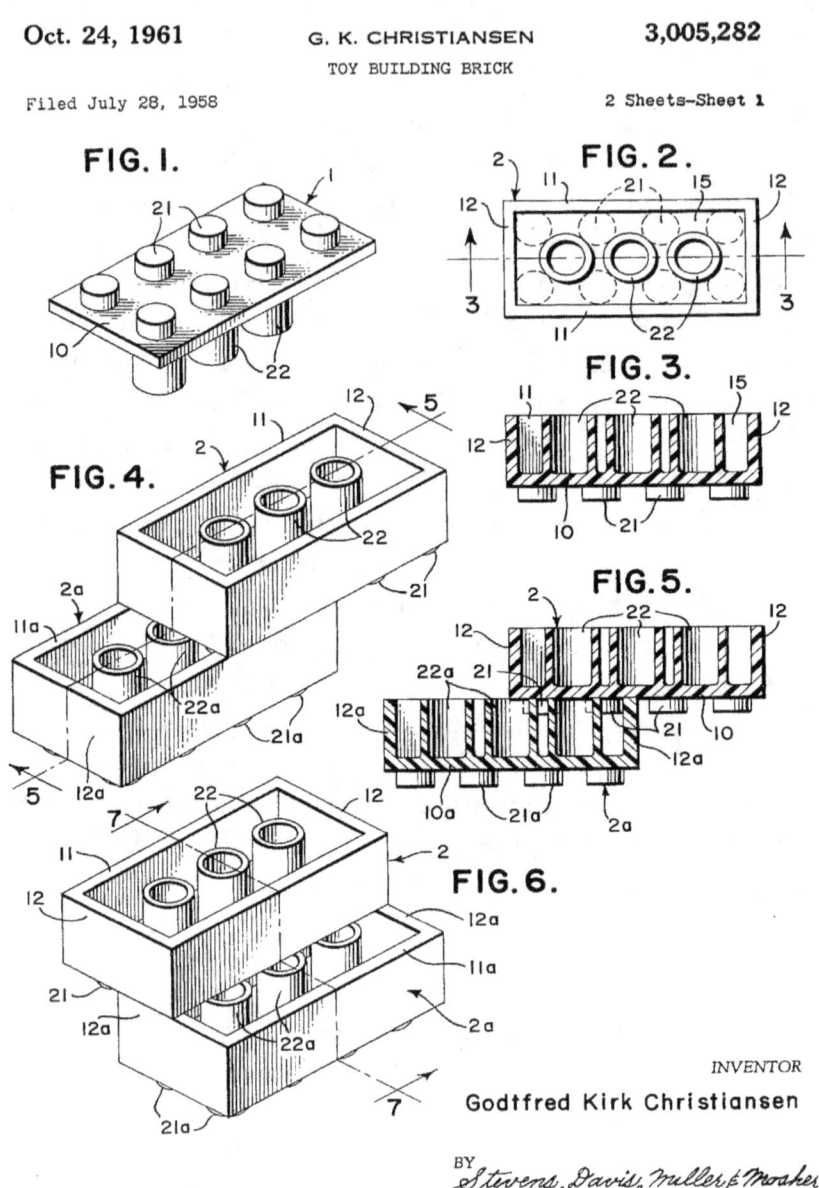

Figure 1.1 Illustration from Godtfred Christiansen's patent application for the stud-and-tube coupling system

Chapter 1 – Building Blocks

Legoland is for young kids, but we had decided to visit it today after Seb had asked about my most recent book. "So, what have you been doing research-wise this year, Dan? Anything in the works now that you've published *The Biblical Clock*?"

"Lots," I replied. "The research for that book really got me traveling down a whole bunch of paths, but the one that most captured my interest was going back to the very beginning. Further back than I went in *The Biblical Clock*."

"The Big Bang?"

"Nope, the *very* beginning—the first instant, when all the building blocks for the universe came into being: time, space, elementary particles, and the forces of nature."

"Whoa, that's big. I've never been able to get my head around it. Are you approaching it as a scientist or as a student of the Bible?"

"Both, like I did when researching the clock book. I'm looking at what science has provided so far in the way of explanations, then at what the Torah says. And I'm finding out some pretty surprising stuff."

"Walk me through some of it, if you feel like it," said Seb. So I did.

≈≈≈

"The scientific approach to explaining how the universe came about at the very beginning starts with identifying *what* needed to come into existence in order for the universe to form via the Big Bang.[2] Clearly, it needed particles of matter, and forces to combine these particles. So how have scientists gone about discovering these?

"Let's think of the universe as like something built out of Lego bricks. Now imagine that you've never heard of or seen Lego, but you come upon this complete Lego structure—a house, say. It doesn't matter what structure you imagine. Say you wanted to figure out how the structure had come together, but you couldn't find the builder or the plans. What would you do?"

"I'd try to take it apart," said Seb.

"Exactly. You'd pull off the roof and take it apart, then you'd take apart the walls. If you couldn't separate some pieces easily, you'd try banging on them to see whether they'd come apart. Eventually, you'd

7

end up discovering that the structure had been made of just a few essential pieces repeatedly put together according to simple rules: snap the back of one onto the front of another or vice versa. The process of taking things apart is a key aspect of the scientific approach to understanding our universe."

"When you say taking things apart, what does that involve?"

"To determine the building blocks of the universe, scientists repeatedly took matter apart until they got to the smallest indivisible parts: the elementary particles, such as electrons. At first, they took substances apart—water, for example, which they split into hydrogen and oxygen. Then they discovered that these elements, and all the other elements in the periodic table, were not the smallest pieces. Experimental scientists began to show that the elements were made of atoms, and atoms were composed of a concentrated nucleus with a positive charge, 'orbited' by electrons. They then discovered that the nucleus contained most of the atom's mass and was composed of protons and neutrons."

"But I remember my science teacher saying that protons and neutrons aren't elementary particles, the fundamental building blocks of matter," said Seb.

"Scientists weren't sure, and there was only one way to find out: smash them to see what resulted. To do this, scientists built particle accelerators to make particles collide at close to the speed of light, then they analyzed what resulted. This is how we now know that a proton is made of three quarks, which scientists believe are one kind of elementary particle."

"I remember learning about how the Higgs boson was discovered using the Large Hadron Collider at CERN in Switzerland. Is that another elementary particle?"

I nodded. "Work at CERN and other particle accelerators has shown that smashing particles doesn't just reveal new ones—it also gives clues about how the particles interact, and it provides insights into the fundamental forces and laws of nature. So, science isn't just discovering the equivalent of all the types of Lego bricks in nature; it's also figuring out how the particles combine and interact. These forces of nature are analogous to the coupling mechanism in Lego bricks."

Chapter 1 – Building Blocks

"So how are the things we look at in nature made up?" Seb asked.

"Well, we've discovered seventeen elementary particles so far, but the things we see every day with the naked eye are made of just three: the electron and two types of quark. Under natural circumstances, quarks always appear as combinations—two of one quark type and one of another quark type, in the case of a proton or a neutron. Protons and neutrons form the nucleus of the atom, with electrons 'orbiting' around it. Atoms then combine to make solid crystals, which in turn make microscopic structures, such as sand, for example, and macroscopic structures, such as sandcastles. Or the atoms combine in less organized structures as liquids or gases."

"Why two types of quark and one electron? And why do they have properties like mass and charge that have specific values? I remember having to memorize some of those numbers for physics class," said Seb, wrinkling his brow.

"The short answer is that we have no idea; these are simply things we've learned by studying matter. We've been able to figure out *what* but not *why*."

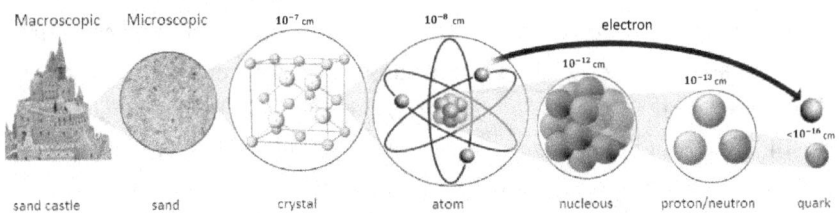

Figure 1.2 The constituents of matter[3]

"So, science has figured out elementary particles and how these make up the things in the world," said Seb. "What about the whole universe—how did it come together?"

"Lego is again useful for understanding this, but first, we need to take a very quick look at the speed of light and how this affects the way scientists study the universe."

The sun was beating down in typical SoCal fashion, so we took advantage of an empty bench in the shade to sit and cool down.

"When we gaze at the night sky, we're not just admiring the twinkling stars and glowing planets—we're actually looking back in time. It takes light a finite amount of time to travel to our eyes, although usually we don't notice this."

"Sometimes," Seb said, "when I'm watching a live interview on TV, when the reporter or the person being interviewed is halfway around the Earth, the transmission is delayed, so there's an awkward pause after a question is asked. Does that have something to do with this?"

"Yes, that's what's going on. The signal is traveling at the speed of light and takes about a second to arrive, leading to those awkward pauses. Our sun is about 150 million kilometers away. Light travels at about 300,000 kilometers per second, so when we look at the sun, we're receiving light that left it about eight minutes earlier. If the sun suddenly went out, we wouldn't know for eight minutes."

"Wow, so that's what you meant about looking back in time when we're stargazing."

I nodded. "When we look at other stars or galaxies, we see light that left perhaps four years ago, a hundred years ago, or a billion years ago. We're seeing every object in the night sky the way it was sometime in the past. So as we look at the universe, it's as though we're viewing snapshots of different parts at different times: Earth as it looks right now, the sun as it looked eight minutes ago, the center of the Milky Way galaxy as it looked 26,000 years ago, and so on. Today, with the Hubble Space Telescope, we can see light that left thirteen billion years ago—not at the beginning of the universe, but close."

"So how does that help scientists figure out the way the universe was formed?"

"Let's go back to the Lego house and imagine that instead of seeing it as it exists now, we were only given pictures of different parts as the house came together: how the side walls looked a quarter of the way through construction, how the front and back doors looked halfway through, and so on. With enough pictures, and by theorizing about how the various parts came together, we might be able to make sense of what the whole structure looked like at all those times and even how it looked when finished."[4]

"OK," said Seb. "So scientists who study the universe can take both of these approaches: taking things apart to discover the

Chapter 1 – Building Blocks

fundamental particles and forces of nature, and looking at the stars to piece together what the universe looked like at various times in its history."

"Yes, they combine the results of these methods with Einstein's theory of general relativity and other theories to reach an understanding of how the universe came to be and what it is today. They encapsulate this knowledge in a cosmological model we know as the Big Bang theory. But we can encounter problems when using these approaches."

"What sort of problems?"

"Sometimes, extrapolating back in time doesn't give us the right answers." A small piece of driftwood was lying in front of the bench, and I bent down to pick it up. "Let's go back to when Lego made toys out of wood, which unfortunately burned extremely well during those three factory fires. Imagine we've never encountered wood before. After taking apart one of these wooden toys, we want to know how the wood came about, so we proceed to take it apart, and we find it's made of 50% carbon, 42% oxygen, 6% hydrogen, 1% nitrogen, and 1% other elements.

"So far, so good. We already know that carbon is made of protons, neutrons, and electrons, and we know that protons and neutrons are made of quarks, and that means we can explain how the toy was made: someone put quarks together, then mixed in electrons to make the key elements such as carbon, then made wood, then made the toy—voilà!"

"But that's not what happened," protested Seb.

"No. The wood grew in a tree that was cut down to make various things, including toy parts. So what went wrong with our reasoning? We followed a good process, but it didn't apply. Why? In this case, wood is an organic compound, and organic compounds usually come from organisms, although sometimes they can be made synthetically, without involving life. In our imaginary scenario, where we had never encountered or heard of wood, we took things apart and followed the synthetic path, unaware of the living organic path."

"Can scientists make the same kind of mistake when they extrapolate back in time with the Big Bang theory?" asked Seb.

"Well, yes and no. It works for getting to *almost* the beginning but not *the* beginning. In this case, they're not fooled into the wrong path, because the equations they use to go back in time actually stop working. They get to a point in time where things were so hot and dense that it's impossible to go further back, because our current theories just don't work in that situation, and there's literally no way on Earth to re-create the conditions at the beginning of the universe and learn about the physics by experiments. In fact, we may never get there, because we would have to go right back to there being no particles, no forces, and even no space and no time. To absolutely nothing."

"Wait," Seb turned. "I get that there'd have to be no particles and no forces, but what are you talking about in terms of time and space? They aren't building blocks."

"If we go back, way back before particles and forces, before anything, before time and space, to absolute nothingness, we need to explain not just how the particles and forces came to be but where time and space came from. To make a Lego structure, we need blocks that click together, but making a universe requires time and space as well as elementary particles and forces. So we need to understand how all this came from nothing at all. Scientists have a good understanding of how the universe came together from an instant after the beginning, up to today, but no idea about what happened in *the first instant*."

"Sounds like we're stuck," said Seb. "How are you working on what happened right at the beginning?"

"Let's go back to the plastic Lego bricks. Even if we understood how the Lego house came together, if we didn't know anything about Lego, we'd have no understanding of how the pieces came to exist. But say we went to Legoland in Billund, next door to the original factory, where the Lego story is told, including with original prototypes. And imagine we had a special invitation to look at the original plans drawn up by Ole and by Godtfred. Then we'd find out how the pieces came to be, why they're shaped as they are, and why they snap together as they do.

"We'd also learn about the Lego System in Play, and that despite variations in the design and purposes of individual pieces over the years, Lego bricks from 1958 still interlock with those made today, and Lego sets for young children are compatible with those made for teenagers; that six pieces of 2x4 bricks can be combined in 915,103,765 ways; that each Lego piece must be manufactured to an exacting degree of precision so that when two pieces are engaged, they fit firmly but are easy to disassemble; that the machines used to manufacture Lego bricks have tolerances as small as 10 micrometers;[5] and so on. Some of this would be pretty much impossible to learn from taking the original Lego structure apart or smashing the individual Lego blocks. But because we were able to look at the inventors' notes, we learned about the blocks' beginning."

Seb laughed. "You're saying we need the inventor's notes for the universe!"

I smiled. "Yup, that's what I'm looking for. Science seems to have reached an impasse, which is why I decided to research biblical sources to look in detail at how the Bible—specifically, the Torah—approaches figuring out how the universe came to be and why it is this way."

"When you say, 'the Torah,' do you mean the Five Books of Moses?"

"Actually, I'm using it in the broader sense, meaning the entire Written Torah—canonized scripture—as well as the Oral Torah, which elaborates on the canonized scripture and contains the mystical tradition, Kabbalah."

"Why the Torah?"

"Because the Torah is the record the Creator left behind, describing His creation."

"Like the Christiansens' original invention plans," Seb smiled.

"Yes, sort of," I replied. "Now, it's important to distinguish the Torah from what we generally call biblical wisdom. According to its own account, the Torah precedes all material existence.[6] The very existence of the cosmos is contingent upon it,[7] and it contains the blueprint that the Creator used to make the universe.[8] A Midrash Rabbah on Genesis says, 'God looked into the Torah and created the world.'"[9]

"That *sounds* easy," said Seb. "Check the Creator's record and learn what happened, just like looking at the Christiansens' invention notes. I'm no expert, Dan, but when I read Genesis, I don't see anything that looks like an inventor's notes."

"That's because the Torah was published first and foremost as a guide for us to live by, like a user's manual. The Ten Commandments are the most famous example from it. You're familiar with a car's user's manual. It describes how to start the car, turn on the headlights, and so forth, but it definitely doesn't tell you how to build the car. For that, you'd need the design manual. In the case of the Torah, we're told that the design information is there, but it's not obvious. We have to delve into the mystical tradition to extract the design manual."[10]

"One sec," said Seb, and he took an iPad out of his backpack. "I just want to pull up the first chapter of Genesis." He tapped and typed, and his eyes scanned the screen.

"OK," I continued, "as a user's manual, Genesis describes the creation of the cosmos in its first chapter. It contains fewer than 800 words, and the description doesn't seem to match what science has established:

> In the beginning God created the heavens and the earth. Now the earth was formless and empty, darkness[11] was over the surface of the deep, and the Spirit of God was hovering over the waters. And God said, "Let there be light," and there was light.

It's so short and general. How do we figure out the laws of physics from this, let alone what happened at the first instant?

"This is where finding out how to extract the design manual that lies within the user's manual comes in," I said.

All of this talking had made me thirsty, so I took out two bottles of water from my own pack and handed one to Seb, then opened mine. After taking a long drink, I held up the bottle and said, "When I first learned some chemistry, I read in a textbook about water and how we can use it—to make ice, make steam, use steam, and so on. I was given a tiny bit of design information: it consists of oxygen and hydrogen."

Chapter 1 – Building Blocks

"Yeah, I learned that in science class last year but didn't find it of much use, because I couldn't envision how those two gases made water," said Seb.

"If you're taking chemistry this year, you'll learn about chemical formulae, the properties of molecules and atoms, and so forth. When I learned about these things, I began to understand a bit more about the design of water.

"In the same way that chemistry books contain words, chemical formulae, and physical chemistry equations, the Torah has many levels and dimensions. For example, the Hebrew letters have numerical values, and each word has a value that's the sum of the letter values. This means that each sentence reads as we would read English but also as a formula."

"I had no idea!" said Seb. "That's fascinating."

"You know about pi (π), right?"

"Sure. It's a mathematical constant expressing the ratio of a circle's circumference to its diameter." Seb closed his eyes and said, "It starts with 3.14159 and goes on to infinity. I used to have it memorized to ten decimal places."

I smiled. "An encyclopedia will tell us that some of the earliest written approximations of pi were from Babylon and Egypt, around 1900–1600 BCE, and were accurate to about 1%, so 3.125. But pi is also contained in the First Book of Kings, one of the books in the Bible, which describes King Solomon making a pool:[12] 'And he cast the pool, ten cubits from edge to edge, round, five cubits deep, and the perimeter surrounding it thirty cubits.'"[13]

"Big deal," shrugged Seb. "So the Bible says in words that pi is approximately 30 divided by 10. Everyone must have known that back then!"

"Yes, but that's kind of like saying water is hydrogen and oxygen, then not knowing anything about how they actually form water. Let's go deeper, though. If we use the simple mathematical values for the Hebrew words in that sentence from the First Book of Kings, we can determine the ratio of circumference to diameter—in other words, pi."

At this point, Seb held up a hand. "Dan, hang on. Since you're getting into specific numbers, do you mind writing them down? It would be easier for me to follow that way." He handed me his iPad.

"Sure." I typed out:

> 3 times the exact ratio of the numerical value of certain words in the Kings quote, or 3 times $111/106 = 3.141509$

"This is very close to today's value of 3.14159265," I said. "But it doesn't stop there. The letters are also pictographs. More complex analysis using the letters' shapes and values allows the calculation of pi from letters and sentences in the Bible. These values come from the design manual rather than by using the scientific method of measuring the diameters and perimeters of circles."

"OK, but pi is a very simple kind of number," protested Seb.

"True, but it's a useful example of what the mystical interpretation of the Torah can determine. The Torah also contains a rich explanation of how the universe came to be. There's the short text in Chapter 1 of Genesis, but the commentaries on and explanations of that core text span hundreds of pages. And all of this literature, which contains the key concepts of the Big Bang and an expanding universe, existed for many hundreds of years when people who studied the natural world primarily believed the universe was eternal and static.

"So in the biblical approach, careful study and analysis of the Torah can reveal how the universe came to be and why it is the way it is. And crucially, since this process gives the design manual that predates the universe, it starts right from the very beginning—the instant in time that scientists may never be able to explain using the scientific method."

"You've definitely piqued my interest," said Seb. "I understand the science approach of looking at what exists and working out where it came from, and the biblical approach of looking at the inventor's notes. Can we go into more detail?"

"Sure. First, though, let's find a quieter place to have our lunch."

Chapter 2

Expanding on Einstein

Above Przemyśl, Austro-Hungary, February 1915

Alexander watched the bombs fall on Przemyśl and smiled. The weather had been excellent on all but one of the missions, and as puffs far below his Muromets bomber showed the payloads detonating on impact, he made several quick notations in the notebook on his lap. He would elaborate on them once he was back on the ground, but the concise version was that his predictions had been confirmed. Alexander looked forward to sending this news to his former mentor, the distinguished mathematician Vladimir Steklov.

Below, the city of Przemyśl was buckling under the combined pressure of aerial and ground assaults. Although the Polish fortress, aided by the Austro-Hungarians, had withstood a 300,000-strong Russian army in the early weeks of the war, the Russians had begun a second siege in October 1914. Reinforced by air support from bombers such as the one Alexander was flying, the siege had held through the long and vicious northern winter, and by March 22, 1915, the Russians would defeat the exhausted, starved town, destroying all its fortifications and capturing 126,000 prisoners. Looking down on the devastation as he turned his bomber east to return to base, Alexander already knew on that afternoon in late February that Przemyśl soon would fall.

For his bravery in the bombing of Przemyśl, he received a military award, after which he was assigned to deliver aeronautics lectures to other pilots. In March 1916, aged just twenty-seven, he was appointed Head of the Central Aeronautical Station, first in Kiev and then, in 1917, in Moscow.

≈≈≈

Born in 1888 to a family of modest means, he had excelled in school, gaining admission to the University of St. Petersburg in 1906. There, he'd come under the influence of Vladimir Steklov, then Chair of Mathematics. Steklov recognized Alexander Friedmann's talent, noting the undergraduate's "outstanding working capacity and knowledge compared with other persons of his age" and ensuring that the financially strapped young man could continue into a master's program.

Friedmann had also been influenced by Austrian mathematician Paul Ehrenfest, who had established a modern physics seminar at the University of St. Petersburg, a dynamic group that gathered to discuss quantum theory, relativity, and statistical mechanics.

In 1913, Friedmann had completed his master's degree and embarked on a career in meteorology, obtaining an appointment at the Aerological Observatory on the outskirts of St. Petersburg. Shortly after war broke out on August 1, 1914, Friedmann joined Russia's volunteer aviation detachment. In addition to carrying out bombing missions, he began theoretically modeling the trajectory of bombs, seeking Steklov's advice about various equations. The older man helped but also tried to discourage Friedmann—"He won't take my advice to keep off bombs and get more into calculations: it would be more productive too," wrote a frustrated Steklov in his diary.

Inadvertently, Friedmann's persistence and scientific pursuits during the war led him to solve equations that were similar to those he was about to encounter relating to the universe. Once Europe had emerged from the chaos of armed conflict, it was once again possible for scientists, their ideas, and their publications to travel, and in some cases, there was much catching up to do.

Einstein's field equations in his theory of general relativity, published in 1915, didn't reach Russia until after the Great War and the revolution, but once they did, Friedmann was swift to engage with Einstein's work. In June of 1922, he published an article titled "On the Curvature of Space," in the respected German journal *Zeitschrift für Physik*. Friedmann showed that the solution to Einstein's equations for the whole universe predicted the universe was expanding from a beginning, posing a major challenge to the prevailing belief that the

Chapter 2 – Expanding on Einstein

universe was static. These solutions were, in effect, the earliest version of the Big Bang theory. However, Einstein initially rejected them, writing to the journal: "The results concerning the non-stationary world, contained in [Friedmann's] work, appear to me suspicious. In reality it turns out that the solution given in it does not satisfy the field equations."

Friedmann was quick to respond, writing directly to Einstein to defend his calculations. Although he sent the letter in December 1922, Einstein remained unaware of its contents until May of the following year. Examining Friedmann's calculations, Einstein realized his own error and promptly wrote to the journal: "In my previous note I criticized [Friedmann's work in 'On the Curvature of Space']. However, my criticism … was based on an error in my calculations. I consider that Mr. Friedmann's results are correct and shed new light."[1]

In 1929, Edwin Hubble at the California Institute of Technology experimentally verified that the universe is indeed expanding from a beginning. But Friedmann was not alive to see this confirmation of the equations he had presented seven years earlier, having succumbed to typhoid fever in the summer of 1926, aged thirty-six.

Einstein went on to publish in 1931 a model of the expanding cosmos that became known as the Friedmann–Einstein universe. In May of that year, Einstein made this model the subject of his second Rhodes Lecture at Oxford University. So significant was this lecture that a blackboard he used to present the ideas is preserved as "Einstein's blackboard" at Oxford's History of Science Museum.

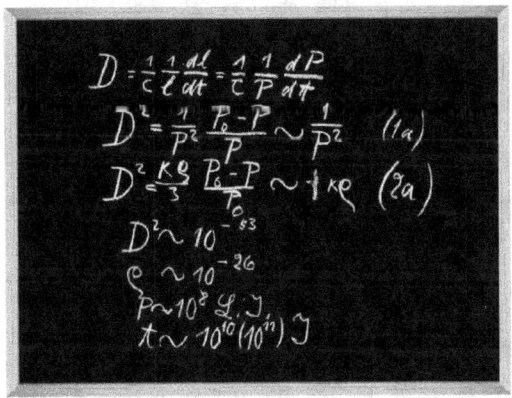

Figure 2.1 Einstein's blackboard

≈≈≈

"That is so cool!" said Seb, reaching for his second sandwich. We'd left Legoland and were eating our lunch at a picnic table in the nearby South Carlsbad State Beach Campground. "But how come I've never heard of Alexander Friedmann before?"

I shrugged and pointed at the sun beating down from a cloudless sky. "If you set off fireworks right now, even really spectacular ones, they'd hardly be visible in the sunshine. That's pretty much what happened to Friedmann alongside Einstein, especially since Friedmann was isolated in Russia and died so young. Who knows what else he might have gone on to do."

"So with Friedmann," said Seb, "scientists accepted the idea that the universe hadn't always existed. It had a beginning. Is that right?"

"Yes, certainly after Hubble published his own observations. And with the Friedmann equations, as they came to be known, it became possible to calculate the universe's age and the trajectory of its expansion. During his short lifetime, the measurements required to ground the equations in our universe's specific parameters weren't available. It took Hubble and others to measure the parameters."

"But I don't get how the equations tell us what happened from the beginning," said Seb.

"OK, it's not obvious because you can't visualize his equations. So let's start with something way simpler: firing a missile in the air."

"What's that got to do with the Friedmann equations?"

"Well, as you now know, he was an expert in the trajectories of bombs and missiles, and it turns out his equations for the universe are very similar to the equation for a missile thrown in the air. So bear with me. Imagine yourself firing a missile straight up. You fire it with all the energy the launcher can deliver, so it goes at its highest speed, and it goes up, slows down, stops, turns around, and falls back to the ground."

Seb's forehead crinkled in concentration. "Yeah, OK, but what's this got to do with the Big Bang?"

"Bear with me a little longer. Imagine you have a modern missile with a motor. You send the missile up into the air, and it rises higher before falling back. With enough help from the motor, the missile would actually escape Earth's gravity and proceed to wander through space, not falling back. These two outcomes—the missile either succumbing to gravity or escaping it—turn out to be the same for the universe as for the missile; even the math is similar."

Seb held up both palms. "Whoa. I don't think I can handle the math."

"It's OK," I smiled, "we don't need to go into that." He looked relieved. "Remember back at Legoland, we talked about the scientific process for understanding the development of the universe. Scientists take things apart, see how they were built, and extrapolate back in time using current physics to figure out what might have happened earlier. They use particle accelerators to smash matter together at high speeds and study what happens during these collisions. This is how we've learned that the universe is built from various particles of matter that interact via various forces of nature. Using computers, they run Friedmann's equations describing the development of the universe, backtracking in time to see what had to exist earlier to lead up to what exists today. The result of these two approaches is the standard Big Bang theory."

"All right. But what does all this show about our real universe?" asked Seb.

"When we run the equations, we trace the history of the universe since it began. Think about video-recording the missile as it's thrown straight up," I said. "Imagine looking at it in midflight on its way up and using the physics we know to go backward, like playing the video backward, to see where the missile came from. Then we run the video forward and see where the missile might end up. Let's run backward first. The current situation is that the universe is expanding."

"So that's when the missile is rising?" asked Seb.

"Exactly. When we use the equations to, in effect, run the video of the universe backward, the universe becomes more compressed, meaning it also becomes hotter and denser. This process continues until the universe is far hotter than the sun and dense beyond our capacity to imagine."

"What happens when we run the equations forward?" he said.

"Well, depending on the current expansion speed and mass of the universe, it either expands, comes to a stop, then begins to collapse, eventually ending up right where it began—like the missile falling back to Earth—or it continues to expand, escaping its own gravity. So far, we don't know enough to be sure of the universe's outcome, but the most likely scenario is expansion forever."

"How much have scientists figured out about what happened after the Big Bang?" said Seb.

"Quite a lot," I answered. "One millionth of a second after the universe came into being, time, space, the elementary particles, and the forces of nature all existed."

"This is when the universe was too hot and dense for us to even imagine?" Seb asked.

"That's right. So we don't know how this first step happened, but let's leave that for later. An instant after this, the universe began cooling down as it expanded, and quarks could now coalesce into particles made of them, specifically protons and neutrons. As cooling continued, the temperature of the universe fell to the point where atomic nuclei could begin forming. Protons and neutrons started to combine into atomic nuclei in the process of nuclear fusion; this is called Big Bang nucleosynthesis. Free neutrons combined with protons to form deuterium, and deuterium rapidly fused into helium-4. This nucleosynthesis only lasted for about seventeen minutes."

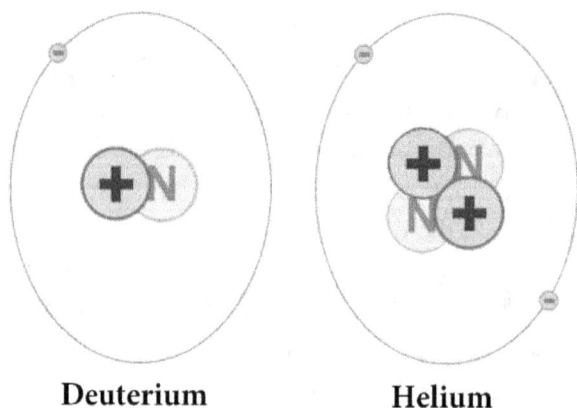

Deuterium **Helium**

Figure 2.2 The composition of deuterium and helium
(N is a neutron, + a proton, and − an electron)

Chapter 2 – Expanding on Einstein

"Seventeen minutes is pretty precise!" Seb looked skeptical.

"You'll have to take my word for it—or I can show you that the equations are precise to fractions of seconds." I said. Seb waved his hands and shook his head. "Scientists have figured out that by then, the temperature and density of the universe would have fallen to the point where nuclear fusion couldn't continue. By that time, all of the neutrons had been incorporated into helium nuclei, leaving about three times more hydrogen than helium-4, by mass, and only trace quantities of other light nuclei. At this point it's been about twenty minutes since the beginning.

"After less than two hundred million years,[2] stars began to form, but they were very different from today's stars. They were short-lived, a hundred times bigger, and a million times brighter, and they emitted much of their radiation in the ultraviolet range."

"What was going on in those stars?" asked Seb.

"Nuclear fusion reactions. Just like in today's stars but back then, with only light gases around, you got these big, bright stars. Inside them, the rest of the elements in the periodic table began to be produced, and over several generations of stars, we got the abundance of elements we find in the periodic table. As these heavier elements were introduced into the universe and into subsequent stars over time, the stars became less bright and longer lived. By about one billion years, they started clustering into galaxies. Then for several billion years, stars and galaxies continued to form, and gravitational attraction gradually pulled nearby galaxies toward each other to form groups, clusters, and superclusters."

"When did our Milky Way galaxy begin to form, and our solar system?" said Seb.

"The galaxy formed at about five billion years and our solar system at about nine billion years. Today, we stand about 13.8 billion years after the beginning."[3]

"And all of this has been figured out in the last eighty or ninety years!" Seb shook his head. "That's amazing."

"Yes. But it gets even more amazing, and mysterious, when dark matter enters the picture," I said. "There turns out to be a lot more matter in the universe, and maybe a lot more energy—which is analogous to our missile with a motor—than what we've been discussing so far. Atoms made of protons, neutrons, and electrons are only 4% of the story."

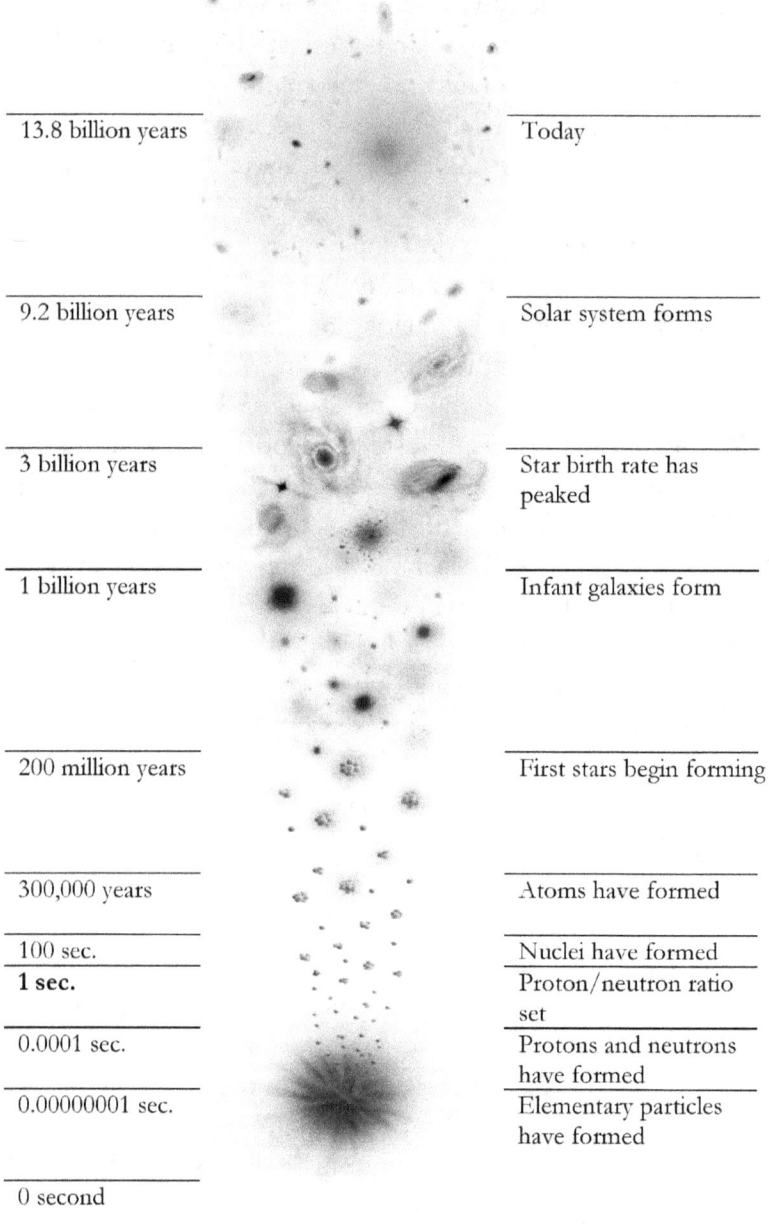

Time	Event
13.8 billion years	Today
9.2 billion years	Solar system forms
3 billion years	Star birth rate has peaked
1 billion years	Infant galaxies form
200 million years	First stars begin forming
300,000 years	Atoms have formed
100 sec.	Nuclei have formed
1 sec.	Proton/neutron ratio set
0.0001 sec.	Protons and neutrons have formed
0.00000001 sec.	Elementary particles have formed
0 second	

Figure 2.3 The cosmic timeline[4]

Chapter 3

Her Dark Materials

Outside Tucson, Arizona, mid-1970s

As darkness inked out the last hints of dusk from the Arizona horizon, Vera glanced at her watch and smiled. She had left Flagstaff four hours earlier, and the observatory at Kitt Peak now beckoned just fifteen minutes' drive away. Kent had said he'd get there before her, but the curtains on his room at the motor hotel had still been closed as she'd backed her Chevy Suburban out of the parking space. Vera felt no compunction about not waking him, knowing he would have done—or rather not done—the same were the situation reversed.

Although they were colleagues and friends and felt tremendous mutual respect, Kent and Vera each believed the other to be less skillful at guiding the telescope, so on these trips, they would race to whichever observatory was the destination. The loser would buy breakfast the following morning. So far, Vera had picked up the tab only once—"because I felt sorry for you that day," she had quipped, and it was true. When it came to functioning on very little sleep, she'd had many years of practice as the mother of four children, whom she had raised while also pursuing graduate studies and then a career in astronomy. Her husband, Robert, was completely supportive, offsetting some of the tremendous pushback she had experienced throughout her education and career.

This, combined with Vera's brilliance and tenacity, was how it came to be that in the late 1960s and early 1970s, she and W. Kent Ford, a gifted scientific instrument maker, had been traveling to various observatories with his remarkable image-tube spectrograph. Used in conjunction with a telescope, it allowed them to amplify starlight to observe astronomical objects previously too dim to see. And for the past several years, they'd been observing and measuring the movement of materials in spiral galaxies, with absolutely startling results.

At first, they had been perplexed and thought either the equipment was malfunctioning or their measurement technique was flawed. But everything repeatedly checked out. After measuring sixty spiral galaxies, they were compelled to confront their findings: the materials furthest out in the galaxies were traveling at the same speed as the materials close to the galaxy centers. Plus, the spiral galaxies were rotating so quickly that they should have flown apart rather than staying together. These findings didn't make sense according to Einstein's general relativity, as the amount of visible matter in the galaxies didn't have enough mass to generate sufficient gravity to keep the galaxies together or the outermost materials traveling that quickly.

"What we're seeing in these spiral galaxies," Vera had said to her husband one evening, "is not what we're getting. My calculations from the data Kent and I have gathered show that these galaxies must contain about five or six times more mass than the visible stars and gases add up to."

Vera recalled that back in 1933, the Swiss astrophysicist Fritz Zwicky at the California Institute of Technology had made similar observations when analyzing galaxies within the Coma cluster, an average of 321 million lights years from Earth. Zwicky had at that time coined the term "dark matter" to account for the additional mass that wasn't visible but must be in these galaxies, but he had been largely ignored.

At Princeton far more recently, in 1973, astrophysicist Jeremiah Ostriker and theoretical cosmologist James Peebles had presented computer simulations that suggested gravitational forces from stars would tear apart spiral galaxies unless the galaxies were surrounded by a halo of dark matter, rather like a bun around a hamburger patty.

Chapter 3 – Her Dark Materials

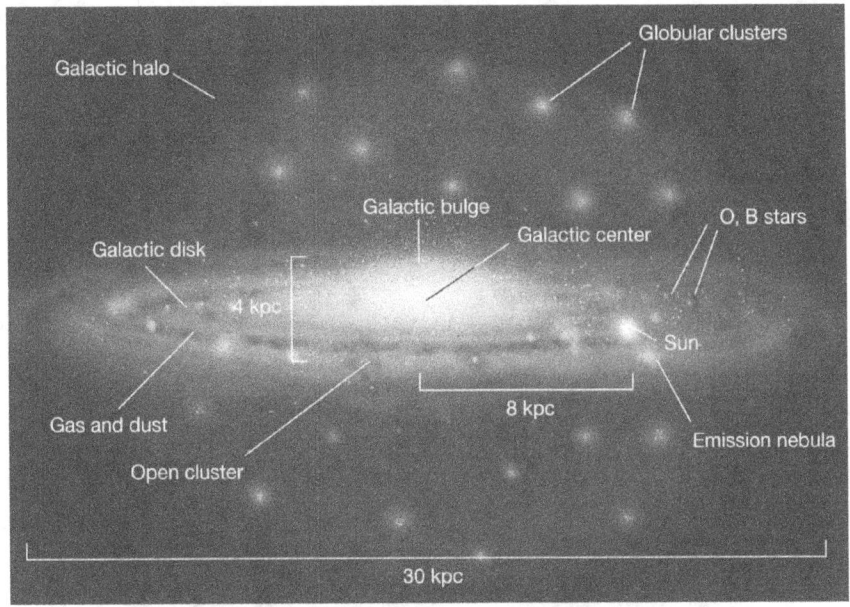

**Figure 3.1 The Milky Way,
enclosed in a galactic halo of dark matter**
(1 kpc is 200 million times our distance to the sun.[1])

Vera believed her data supported the theory of dark matter. Her work routinely met with more skepticism than that of her almost entirely male colleagues; even after receiving her PhD from Georgetown University at age twenty-six, she had wondered, "Will I ever really be an astronomer?" But the amount of data she had gathered from sixty different galaxies was too compelling to be ignored. It had become clear from her calculations that about eighty-five percent of the mass in galaxies is invisible.

In a later interview, Rubin commented, "Nobody ever told us all matter radiated [light]. We just assumed it did." Or as Dennis the Menace said in a cartoon she liked to use: "Lots of things are invisible, but we don't know how many because we can't see them."

Figure 3.2 Vera Rubin standing by the Lowell Observatory 72-inch telescope, Flagstaff, Arizona, 1965

≈≈≈

After finishing our lunch, Seb and I had left some crusts for the seagulls and were walking along the seashore.

"So the matter we see through our telescopes—what we call normal or visible matter—is only a fraction of what is needed for everything to work," I concluded.

Seb was frowning in concentration. "I think I get the stuff about galaxies and rotation speeds, but I'm not quite sure…"

Chapter 3 – Her Dark Materials

"That's not surprising," I said, "because it's not what intuitively makes sense based on our everyday experiences. I'll try to clarify.

"The stars inside a galaxy, such as the Milky Way, rotate around the galactic center. By measuring the rotation speed of these stars, we can infer how much matter must be inside their orbit to keep them going around. Think of yourself spinning a kid in a circle while holding onto their arms. The faster you spin the kid, the stronger the force you need to keep hold of them. The same goes for stars. A certain amount of gravity is required to hold them as they go around at a particular speed—not enough gravity and they fly off, like a kid would if you let go of his hands."

"That makes it a lot easier to understand, Dan. So when Vera Rubin looked at galaxies, she found that the visible matter she saw wasn't nearly enough to hold the stars in their orbits around the galactic center?"

"Exactly. She reasoned that each galaxy must be enclosed in a halo of other matter that amounts to much more than the matter we see."

"What is this stuff?" asked Seb.

"We don't know. Like visible matter, dark matter is also found between galaxies and in space in general. By now, astronomers have measured it in thousands of galaxies and spent billions of dollars trying to detect it. In the process, they've determined that it's many times more abundant than visible matter. Scientists have also discovered that dark matter forms the scaffolding on which all visible matter is pinned."

"An invisible scaffolding? That's going to make it tough to study," Seb grimaced.

I nodded. "Yes and no, because the visible matter is 'pinned' to this dark matter. When you look at a picture of a section of the universe, you see bright regions containing visible matter, which look like filaments, and dark regions, which look like voids. The dark matter must be attracting the smaller amount of visible matter, so in a way, we see where the dark matter is more concentrated—right where the visible matter is. The large-scale structure of the universe is made up of these voids and filaments."

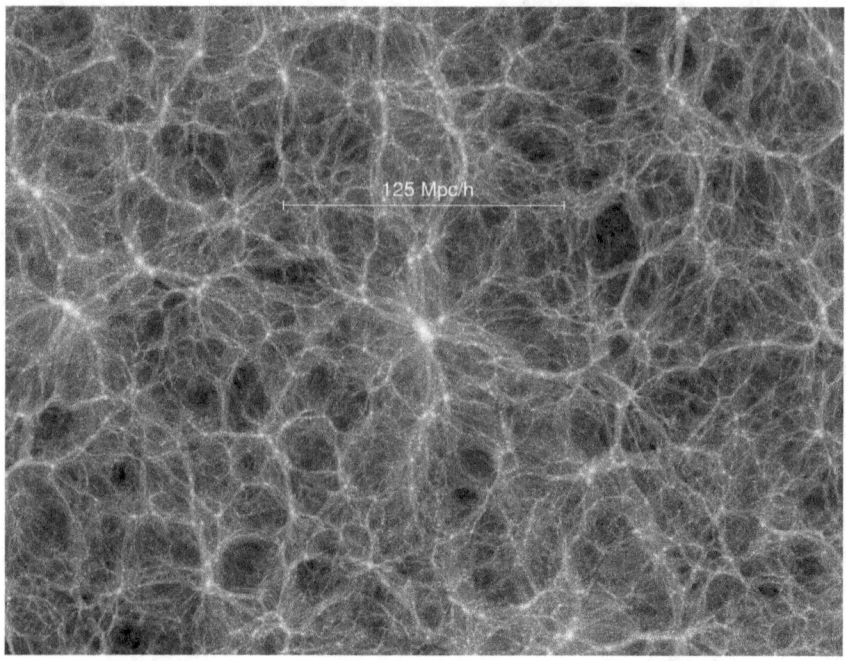

Figure 3.3 Picture of a section of the universe

The scale shown of 125 Mpc/h is about 45 trillion times our distance to the sun.[2] We see filaments (bright lines), voids (dark areas), and, in the center of the image, a huge galaxy supercluster (bright spot). Spread over the whole image are clusters of galaxies and other superclusters. The entire image from left to right measures approximately 1.7 billion light years, about 20,000 times the size of the Milky Way but a mere sliver of the universe.

"What are the filaments made of?" asked Seb.

"Superclusters, clusters, galaxy groups, and galaxies. Galaxies, as you already know, are made up of stars and their constituents, our own solar system being one of these."

Seb stopped abruptly, hands on his hips. "OK, but what does dark matter have to do with the first instant of the universe?"

"Well, it's crucial to have a complete understanding of what the universe is made of, even if some of our understanding is still theoretical because we haven't found stuff yet—like dark matter. Just a few decades ago, the Higgs boson still existed only in theory, but eventually, scientists did prove its existence."

"Fair enough," said Seb.

Chapter 3 – Her Dark Materials

"Once there was a general scientific consensus to assume the existence of dark matter, so that all the matter in the universe could be accounted for, scientists applied Friedmann's equations to the universe, this time with the right amount of matter. The equations predicted an expanding universe but one whose expansion was slowing down due to the universe's own gravitational attraction. And that prediction could now be compared to measurements of the universe's expansion. In 1998, the expansion of the universe was measured more precisely, and it's been further refined since then. But a problem appeared."[3]

"Why did I know you were going to say that?" Seb said wryly.

I smiled. "The universe is a mysterious place. Science has discovered that the universe's expansion is now speeding up, not slowing down. In other words, the universe is acting like a missile with a rocket that accelerates the missile even while Earth's gravity pulls on it. Although the universe's expansion did slow down for the first seven billion years or so, it's been accelerating ever since!"

"Really?" Seb's eyebrows were raised.

"Yup. Many attempts have been made to explain this acceleration of the universe's expansion. Some people have even suggested that Einstein's equations aren't right. But so far, the best explanation for the current data is that a mysterious energy is causing the expansion. In other words, the universe is not acting like an ordinary missile but like a missile with a motor that speeds it up even while gravity pulls on it. Imagine that we launch this missile. It begins to slow down because gravity is very strong, but at some point, the thrust from its motor becomes stronger than the weakening gravity, and the missile accelerates from then on. This is exactly the behavior of the universe."

"Wow. I learned that gravity always attracts," said Seb.

"That's correct. But this mysterious energy that repels matter was always there. In the early universe, the normal gravitational attraction from all of the matter was dominant, because the universe was smaller and so all matter was closer together. This slowed the universe's expansion. Today, though, in the much larger universe where matter is more spread out, the repelling energy dominates, so it accelerates the expansion. Scientists have dubbed it 'dark energy.' We know even less about this than we do about dark matter."

"But you said they discovered the universe was accelerating in 1998. How can we not have figured out something about dark energy since then?"

"In my research, I've come across something that the American astrophysicist Adam Riess wrote about this in 2017. He, along with Saul Perlmutter and Brian P. Schmidt, received the Nobel Prize in Physics for providing evidence that the universe's expansion is accelerating. Let me find it." I stopped and pulled my iPad from my knapsack, then tapped on it to open a file. "Here it is:

> after studying the situation for nearly two decades, the physical nature of dark energy remains almost as elusive today as it was 19 years ago. In fact, the latest observations only seem to further complicate the picture by showing hints of disagreement with the leading theories. ... [W]ithin the next decade we hope to begin to comprehend the nature of cosmic acceleration, or resign ourselves to leaving some mysteries unsolved indefinitely.[4]

"So," I raised my shoulders, "there's still a long way to go."

"What does he mean by disagreements with leading theories?" asked Seb.

"Well, it turns out there are two ways to calculate the acceleration.[5] We can look at nearby, well-understood objects, then observe those same types of objects in more distant locations, then infer their distances, then use properties we observe at those distances to go even farther. In this way, we reconstruct the expansion rate of the universe by looking from here back to the beginning. Or we can use the light left over from the Big Bang, which is called the cosmic microwave background. By examining this, we can see how those signals changed over time as the universe expanded. In this method, we're looking from the beginning up to now."

"And?"

Chapter 3 – Her Dark Materials

"Logic dictates that if our theories are correct, we should get the same answer from both methods. But they actually yield different answers, which implies we don't know exactly what we're doing!"

"If it does exist," said Seb, "how much of it is there in the universe?"

I glanced at my iPad and tapped a few times to open a data spreadsheet. "Assuming that the standard model of cosmology—the Big Bang—is correct, the best current measurements[6] indicate that dark energy contributes 68.3% of the total energy in the present-day observable universe, dark matter 26.8%, and ordinary visible matter 4.9%. Of the ordinary matter, 4% is hydrogen and helium, and 0.9% is everything else—the stuff you actually see in the night sky! Other components, such as photons, contribute a very small amount. Dark energy has a very low density, much less than the ordinary matter or dark matter within galaxies. But it dominates the universe because it's uniform across all of space. This means it's possible that dark energy will cause the universe to continue expanding forever."

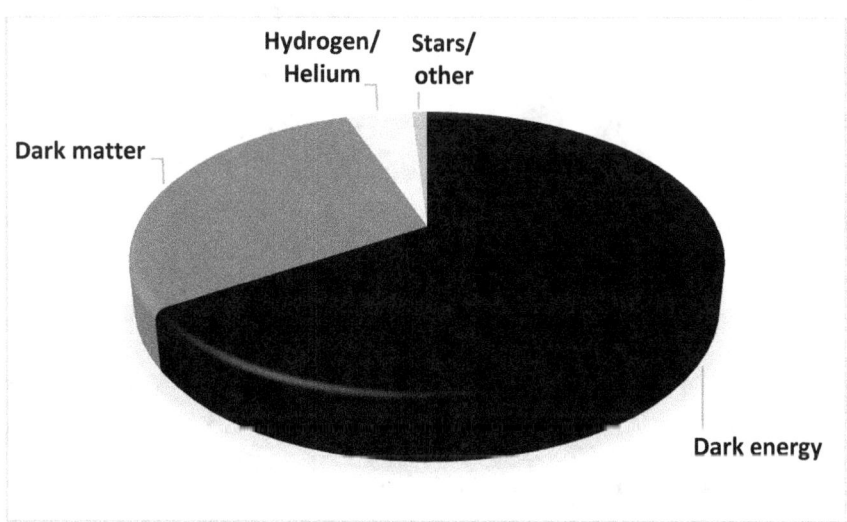

Figure 3.4 Constituents of the universe

"But apart from the dark stuff that we'll hopefully discover," said Seb, "the Big Bang theory is really successful, right?"

"Yes, for explaining how the universe developed from a very hot, dense state about one millionth of a second after time zero to what it is today. But there's a big catch: it says nothing about the beginning."

"Then that's where we go with this next?" asked Seb.

"Not quite yet. First, we need to look at what the universe is made of—the particles and the forces of nature—and what is known about time and space."

Chapter 4

Stay Tuned

Bell Laboratories, Murray Hill, New Jersey, early May 1964

"Pigeons?"

"Pigeons. Nesting."

Taking a few sips of coffee, Robert shook his head and chuckled. The staff cafeteria at Bell Labs was nearly empty in the mid-afternoon, and his laughter made a few of the patrons turn their heads in more than mild interest. Professor Dicke, who worked in Princeton's Physics Department, was by no means a frequent visitor at the Bell Labs; many of the staff there were familiar with using one or more of his inventions, including the "Dicke" microwave radiometer for measuring the energy emitted at microwave wavelengths, but had never seen him in person.

Arno leaned forward, smiling, and said, "Can you imagine how embarrassing it would have been if we hadn't noticed them?"

"Embarrassing?" snorted Bob. "We'd have been the biggest laughingstocks from here to Houston and back. And I can only imagine what you," he nodded and winked at Robert, "and your pals forty miles down the road from us would have had to say. We'd never have lived it down."

"But you did notice," said Robert, "and, well, here we are. It's tremendously exciting. Congratulations."

"It was pure luck that we found it before you and your team did, Robert," said Arno. "We weren't even looking for the stuff, just trying to detect the faint radio waves coming off the Echo balloon satellites."

Bob nodded. "Yeah. We got rid of the interference from radar and radio broadcasts first, then cooled the receiver. Still got plenty of interference. Tested all the equipment again, but that didn't help either.

So we went out to have a look, and there were a couple of pigeons who'd built their nest in the antenna! Once we'd removed that and the bit of a mess they'd made, we figured we'd finally gotten to the bottom of the issue."

"So what first clued you in to the possibility of it being cosmic microwave background radiation?" asked Robert.

"Well, for one thing, it was coming from every part of the sky, day and night. I think we both knew that back in '46, you'd already predicted that something like this existed, some thermal echo from the Big Bang that would have that characteristic."

Arno nodded. "We were both kids back then, obviously. But I know that for me, at any rate, I became familiar with your idea soon after I learned about the Big Bang theory. Then Bernie Burke, at MIT, heard that we'd found this constant, pervasive buzzing we couldn't figure out, and he told me about the paper Jim Peebles was getting ready for publication, on the possibility of finding radiation left over from the 'primordial fireball.' That's when we got in touch with you, and you got Jim's permission to send us his manuscript."

Robert sat back. "Jim was an excellent doctoral student, and we've kept working together since he got his degree. As you know, he and David Wilkinson and I were working on something to detect the CMB at the same time as you two stumbled on it. No offense."

"None taken, stumbled is absolutely right," said Bob.

"So what do you have in mind now?" asked Robert.

"That's why we asked you to meet with us today," Bob replied. "Obviously, Arno and I want to publish the initial findings as soon as possible, and *Astrophysical Journal Letters* has already said they'll keep space for us in the next issue if we can submit in the next two weeks. But we want to know whether you'd be interested in publishing a letter in the same issue, since you and your group had already made the theoretical inroads and would have detected the microwaves first if we hadn't serendipitously been working on our unrelated project."

"Good thing it was only a pair of pigeons," said Robert, smiling archly. "If there'd been more of them, they might've done sufficient damage to delay things long enough for us to beat you to the punch!"

≈≈≈

Chapter 4 – Stay Tuned

"And that," I said, looking at Seb, "is how Robert Dicke *didn't* get a Nobel Prize for his work on cosmic microwave background radiation. Which is unfortunate, because in addition to that, he contributed hugely to physics and cosmology in many other ways, including the next thing we should talk about in relation to the first instant."

We had stopped for a minute to buy some juice from a beach-side stand, and I'd been telling him this anecdote while we rehydrated in the shade of a massive palm tree.

"Cosmic microwave background radiation," said Seb. "I hope you're going to enlighten me on what that is."

"Don't worry, I will," I replied, smiling.

"Good. So the guy who'd come up with the idea back in 1946, and then was specifically working on a way to detect this radiation, got scooped by two other scientists down the road who weren't even looking for the stuff? And they got a Nobel Prize but he didn't?"

"That's right. The paths of scientific discovery can be pretty random sometimes," I said, shrugging.

"You're not kidding! He must've been pretty frustrated about not getting the prize."

"Well, when interviewed twenty years later—which was seven years after the prize went to the other guys—he was pretty philosophical about it. And he seems to have been a humble man. Later in life, when he was admired for having more than fifty patents, he was prouder of the fact that his father, who'd been an electrical engineer and inventor, had even more patents than that."

"That's a cool attitude." Seb smiled.

"I agree. But I'm actually bringing him up because he was a proponent of the idea of a 'fine-tuned universe,' and that's a puzzle we need to cover as we build up the correct picture of the universe just a little after it got going."

"All right," said Seb. "Tell me about this puzzle."

"Remember that the universe developed from a very hot, dense state about a millionth of a second after time zero to what it is today. If we're to run the Big Bang theory from this very early time, we need to know a whole pile of different parameters that go into the model."

Seb took a long drink, then said, "A whole pile. Sounds daunting."

"Maybe, but it actually isn't. Think of Lego again. To build a Lego structure, we need to have all the necessary bricks and know how they snap together. When it comes to the universe, with its vastly more complex structure, we need at least twenty-six parameters to describe the particles in it and all the forces and interactions that occur between them."

"Still sounds scary." His face wore a look of exaggerated fear.

"Some of them you already know. For example, there are the masses of protons, neutrons,[a] and electrons, and the charges of protons and electrons. You're also familiar with the strength of gravity, of the nuclear force that binds nuclei together, and of the electromagnetic force that draws protons and electrons together."

"What about dark matter and dark energy?"

I nodded. "Yes, the amounts of dark matter and dark energy in the universe are two more of the parameters. And it turns out that all of these had to be precisely the values that we've measured for them. This is known as the fine-tuning problem: the fact that the parameters are incredibly fine-tuned for our universe to exist."

"What would have happened if they weren't fine-tuned?"

I rubbed my face, then smiled. "That's a really good question that a lot of very smart people have been pondering for quite a while."

Seb pulled a wry face. "OK, please explain this fine-tuning stuff."

"Remember earlier, we were talking about launching a missile straight up. It rises, slows down, stops, and falls back to the ground. Imagine launching it so that it falls back onto exactly the same spot—right on the launcher."

"That would be virtually impossible," protested Seb. "You'd have to be so precise."

"I agree. Next, you decide to put the launcher in a wide, shallow depression. Up goes the missile, but this time, it falls within the depression and rolls back to the launcher, even if you weren't very precise about launching it so it would come back to you."

---

[a] Protons and neutrons are not elementary but made of three quarks each.

Chapter 4 – Stay Tuned

"I think I see where you're going with this," Seb interjected. "In the first case, unless the launch angle and other parameters are incredibly fine-tuned to a precise value, there's no way the missile will land back in the same place. But in the second case, as long as the launcher is at the lowest part of the depression, the missile will end up where it started."

"Exactly."

"So far, so good. I'm not lost yet," he grinned.

"Great! Well, we've discovered that getting the universe started is like launching the missile: unless the launch is incredibly precise, a miss is as good as a mile. If they aren't these precise, fine-tuned values, we don't get a universe—not ours, anyway. When I was studying science at your age, some hoped it would be like throwing the missile from a hole: if the initial conditions had roughly the right values, everything would work, no fine-tuning required."

"Here comes another 'but'!"

I nodded. "Mm-hmm. But after forty years of discovery, the opposite has turned out to be true. The more we study the universe, the more we realize how fine-tuned the parameters have to be for it to exist."

"How come?"

I shrugged. "We have no *proven* explanation for the fine-tuning, and no explanation at all for the values of the parameters that we measure or for the relations between the values."

"Like the ratio of the mass of a proton to an electron?"

"Like that."

"What would happen if the parameters were different?" he asked, picking up a driftwood stick and whacking a small stone into the surf.

"We'd need a computer to look at all of the possible outcomes if multiple parameters were different."

"Fair enough. How about just one by one?"

"Sure. I was reading about this a few days before we flew down here, actually.[1] Take an atom, which, as you know, has a nucleus surrounded by one or more electrons. The nucleus is made up of protons and neutrons, and protons are just 0.2% lighter than neutrons. If we reversed this, so neutrons were 0.2% lighter than protons, then atoms would collapse, because electrons would combine with protons to make neutrons, and the universe would contain just chunks of neutron matter but no life."

"Seems like such a tiny difference. That really is fine-tuning," said Seb.

"The same goes for the force that binds nuclei together. If it had been even 1% stronger, all of the hydrogen would have combined in the first minute of the universe, leaving no hydrogen for stars to burn and none to generate water. Making it 1% weaker would have prevented heavier elements[b] from forming, so we'd have no planets or people. And there's the ratio of the strength of the electromagnetic force to the strength of the gravitational force. If it were even a little bit larger, all stars would have been at least 40% more massive than the sun, so they'd have burned too briefly and unevenly for life to emerge. Make the electromagnetic force even a little bit smaller and all stars would have been at least 20% less massive than the sun, so they wouldn't have produced the heavy elements that we need for life."

"What if something about dark energy was a little bit different, like the amount of it in the universe?"

"Fine-tuning applies to that, too. Even a slightly larger amount of dark energy would have caused space to expand so fast that galaxies wouldn't have formed. There's also what's often called the roughness of matter problem or structure problem."

"I'm assuming that isn't roughness in the usual sense of the word."

"You're right. It describes the small structure of matter, which had to be a particular way from the very beginning of the universe. Matter is, and was back then too, quite uniform but with a bit of this roughness—think of it as like ripples in water. Because it was like this, gravity brought areas of slightly higher density together, which eventually formed into stars and galaxies. If the roughness had been ten times larger—if the 'ripples' had been more pronounced—then the areas of higher density would quickly have attracted each other so

---

[b] The early universe consisted mostly of hydrogen and helium, then stars formed and, in their cores, lighter elements began to fuse into heavier ones, giving off energy in the process. This created the rest of the elements in the periodic table required for us to exist.

Chapter 4 – Stay Tuned

strongly that only huge black holes would have formed, but no galaxies. On the other hand, if the roughness had been ten times smaller, matter would have been so smooth that the slightly denser parts wouldn't have been dense enough to attract each other fast enough as the universe expanded, so no structure would have formed. The universe would have been just hydrogen and helium forever, no planets or galaxies."

"Ten times sounds like a lot, not fine-tuning," said Seb.

"Actually, when you think of the cosmos, ten times *is* fine-tuning. In cosmology, forty zeros is the kind of range scientists usually deal in, so one zero is small. But even if we postulate a variation in the roughness that's smaller than ten times, we'd still not get our universe. In-between things would form, but nothing like what we see, so it would be even more improbable that something like Earth would ever happen. Cosmologists had been hoping that it would be easy to end up with our universe naturally."

"Like launching a missile in a big, wide depression," said Seb.

"Right. They thought that even if the fundamental parameters had been different, the same universe would have formed. But what they've actually found is that the universe's formation was highly improbable. And the problems don't end there."

"Wait." Seb stopped, smiling, and held up his hand. "Just tell me that at some point soon, we'll start discussing solutions to the problems!"

"For sure. Just bear with me a little longer, because we have to consider the issue of a self-assembling universe." Seb gave a thumbs-up.

"Unlike with the Lego example, where we'd have all the pieces laid out and an intelligent being like us would put them together, the Big Bang theory postulates that the universe self-assembled. Meaning that once all the particles and forces existed, and the universe was very dense and hot but expanding and cooling, the particles and forces interacted naturally to generate the composition of the universe. By composition I mean what we can see and measure about our universe, such as:

- the same number of galaxies, of the same mass, clustered together in the same ways
- the same ratios of the elements in the periodic table found in the universe today
- the same number of stars and planets with the same mass distribution"

"How would that self-assembly have happened?" asked Seb.

"That's a key question. Imagine looking at a stretch of bare soil, then watching how it changes after a lot of rainfall. Everything is rearranged by the simple action of gravity on water. The same goes for the universe. The idea is that forces—primarily gravity in terms of the big structure—acting on all the constituents of the universe led to what we observe today."

"That sounds like self-assembly."

"It does, but as we saw, when we discussed the structure problem, the roughness or structure of the early universe had to be tuned for this to happen. And that's just one of three problems demanding that the constituents of the universe all needed to be arranged in a certain fine-tuned way."

"How come?"

"The next reason is called the horizon problem. When we observe the universe, no matter what direction we look in, we see basically the same things—the universe is very homogeneous. If the universe started with even very slightly different temperatures in different places, we wouldn't see homogeneity. So there had to be a mechanism early on to even out the temperature, a mechanism to have everything in contact long enough to reach equilibrium. This would be like putting cold and warm water together and eventually getting water of the same temperature. Yet parts of the universe have never come in contact with one another and therefore can't have influenced one another."

"How do we know they've never come in contact?"

"It's a bit complicated to calculate, but we've figured out that in the early universe, as expansion got under way, there wasn't enough time for everything that we now see to have come into contact."

"So," said Seb, "there are all these places in outer space that can never have interacted with each other, but they still have basically the same things in them?"

"Exactly. And unless they originated from the same initial conditions, it's extremely unlikely they'd be essentially similar. Even a tiny difference in the initial temperature, for example, would have led to different residual electromagnetic radiation from the early stages of the universe, what scientists call the cosmic microwave background."

"That's the stuff Dicke and the other guys found fifty-something years ago, right?"

I nodded. "But it's the same no matter where in the sky we measure it. So unless there was a mechanism that evened out the temperature between different areas, and that did this more quickly than the universe was expanding, we can't explain the homogeneity."

"I get it."

"Good, because now comes another tricky conundrum, called the flatness problem. Science has determined that the density of matter and energy in the universe is at a precise critical value that makes the universe 'flat' geometrically. In high school geometry class, you might have drawn a triangle on a flat piece of paper and seen that all the angles added up to 180 degrees. If you drew a triangle on a sphere, the angles would not add up to 180 degrees. We don't know what shape the universe is, but we can measure how close to flat it is—in other words, how close the sum of the triangle's angles is to 180. And it's very flat![2] The universe has expanded dramatically since its beginning. So if the density of its matter and energy right now have to be precise (like drawing on a flat sheet of paper), imagine how much more precise it had to be at the beginning to achieve that flat structure today. It's hard to conceive how this could have been the case just randomly."

"OK, that makes sense. And the third problem?"

"We've actually already discussed it: the structure problem. Resolving the horizon problem demands a very smooth universe at a large scale. But if it had been totally smooth, no structure would have developed. At a large scale, the universe has some regions with relatively few galaxies in comparison with other regions that have many more. As we saw, it's easy to understand how this structure came

43

about. The volumes of space in the early universe that had slightly greater mass attracted, via gravity, the matter from adjacent volumes and became denser. Gravity then caused the matter in these volumes to collapse and form stars, galaxies, and galaxy clusters. Meanwhile, the volumes losing mass eventually ended up as giant voids. The universe's finer structure came about in the same way." I glanced at him. "You still with me?" He nodded and smiled.

"Good. Science has established that the visible universe consists of many distinct regions. We'd expect huge initial variations in matter's density across very large distances, and much smaller initial variations on a small scale. But observations indicate that the size of the initial variations at all of these scales had to have been roughly the same order of magnitude."

"So that means they were really fine-tuned?"

"Extremely."

"What's the solution to all this?"

I looked at my watch and said, "Why don't we get back to the car and find somewhere to surf first. Riding the waves for a couple of hours will give both of us a break before we dive into that!"

Chapter 5

Galloping Inflation

College Park, Maryland, March 1980

"An exciting opportunity lies just ahead if you are not too timid." Alan leaned back in his chair and smiled; one eyebrow raised. Not one for taking fortune cookie messages seriously, or horoscopes, for that matter—he was quite sure the stars and planets were absolutely indifferent to humanity—he nonetheless acknowledged that the words printed crookedly on this slip of paper seemed... apt, in his present circumstances.

Arranged around the large, circular table, which sported a disposable tablecloth now liberally spotted with various sauces, were several other physicists, most of them from the University of Maryland. This particular restaurant was popular with postdocs and profs, and although Alan Guth had probably eaten more Chinese meals during the past six weeks than in the previous year, he'd still enjoyed the food and the company. U of Maryland was his last stop on a lecture tour of universities that had expressed interest in hiring him, and the trip had yielded no fewer than eight offers of faculty positions.

The contrast with his life even six months ago was startling. At that point, he had been one of numerous baby boomers with science PhDs and no immediate faculty prospects—"all gowned up with nowhere to go," as a friend at a job fair a couple of years ago had quipped. There were too few assistant professorships to go around. Married and with a two-year-old son, Alan counted himself lucky to be in his ninth year as a postdoc, rather than teaching in a high school or working in an entirely unrelated area.

Then, one evening in early December of 1979, things in his professional life had started to undergo a sea change. Recounting the experience many years later, Alan—by that time, the Victor F.

Weisskopf Professor of Physics at the Massachusetts Institute of Technology—would call it "serendipity, in that what I found that night was not really what I was working on but in fact gave an answer to one of the important fundamental problems in cosmology."

He was employed at the Stanford Linear Accelerator Center, better known by its snappy and rather ironic acronym, SLAC. Neither Alan nor his colleagues could be called slackers. Although Professor Leonard Susskind, listening to Alan's new theory a few days later, would say to him, "You know, the most amazing thing is that they pay us for this," the truth was that they all put in long hours, and having a "work–life balance" sounded even more fantastically theoretical than their musings about fundamental particles.

That night, Alan worked on his calculations until 1:15 am; a meticulous diarist, he had noted this in his entry for December 6. He and Henry Tye, whom he'd met the previous year at Cornell, were co-authoring a paper about magnetic monopoles. It was an "oddball project" that had them figuring out how many of these magnetic monopoles—which to this day, no one has ever actually found—would have been produced in the Big Bang if Grand Unified Theories (ingloriously called GUTs) were right.

They'd already figured out that assuming GUTs and regular cosmology were correct, so many magnetic monopoles would have been produced that "the universe would be swimming with them." So what needed to be changed in the theory to get around this magnetic monopole glut and make GUTs consistent with cosmology?

It was accepted that in the universe's first few seconds, as it expanded, super-cooling had taken place. Henry and Alan had been assuming that while this cooling was going on, the universe would have continued expanding in the same way. Henry then suggested they reexamine this assumption. So that's what Alan was sitting at his desk doing.

Going through the equations, he realized that actually, this super-cooling would have triggered very rapid exponential expansion. As Alan checked and rechecked his calculations, he saw that this exponential expansion could solve a fundamental hitch in cosmology.

Chapter 5 – Galloping Inflation

"By chance, all of the pieces had fallen into my lap, and I just needed to put them together."

The hitch was the flatness problem. Cosmologists had figured out that if the expansion during the first second of the universe had been faster or slower by just a tiny bit—as little as 10^{-15}—then the universe would either have flown apart or re-collapsed on itself.[1] In either case, nothing in our present universe would have existed.

What Alan realized was that this exponential expansion that solved the magnetic monopole problem would also have made the universe get flatter and flatter as it got bigger and bigger, thereby solving the flatness problem.

A few weeks later—"December 1979 was my lucky month," Alan wrote in his 1997 book *The Inflationary Universe*—he was having lunch in the SLAC cafeteria, listening to some other physicists discuss a recent paper about the horizon problem. As a theoretical physicist rather than a cosmologist, Alan had until then been unaware of this long-standing issue. That afternoon, though, he thought, "Eureka! The exponential expansion of inflation would obliterate this problem, too." His theory posited an initial size for the universe so miniscule that the uniformity of temperature we now observe in places vastly distant from one another would have been achieved without difficulty just before the massive expansion occurred.

These ideas were so potentially big that Alan wanted to share them as quickly as possible. Publishing them in a journal article would be essential, of course, but still too time-consuming, especially because he always took ages to write a paper. "I'll have to give a talk," he decided.

So on January 23, 1980, he débuted the "inflation" concept at SLAC, in a talk titled: "10^{-35} Seconds After the Big Bang." News of his ideas spread remarkably quickly, and by the third week of February, he had embarked on his five-week lecture tour of nine universities and Fermilab, the world's most powerful particle accelerator facility at that time.

By the evening that Alan read the amusingly apt fortune cookie's prediction, he knew that his days as a peripatetic theoretical physicist were over. Opportunity was knocking, and he certainly wasn't too timid to answer.

47

≈≈≈

"What a cool story," said Seb, adjusting the air conditioning dial in the rental car. We were on our way back to the hotel after some great surfing, and the traffic was snarled due to an accident, so I'd begun telling him about Alan Guth's breakthrough theory of cosmic inflation. "Imagine solving such big problems. That's got to be what every scientist dreams of doing!" Crossing his arms, he continued, "I'm not a hundred percent sure that I get the inflation idea. Do you mind explaining it?"

"Sure," I said. "Here's the issue. Somehow, after the universe started, it had to quickly distribute its properties to yield a universe that expanded to what we see today. Let's start with the flatness problem: no matter what curvature the universe had at the beginning, it had to become 'flat' everywhere. How do you take a curved surface and flatten it? Think about a small balloon with a curved surface. Now, imagine inflating it to millions and millions of times its original size. At this point, any small piece of the balloon's surface will now look flat. The theory of inflation basically works like this balloon. The idea is that the 'patch' out of which our space–time emerged was about ten billion times smaller than the diameter of the nucleus of an atom. By the time inflation ended, about 10^{-33} seconds or so following the Big Bang, this little patch had stretched until it became a few millimeters to centimeters in size. After that, it expanded according to the standard Big Bang theory and has continued expanding to the present day."

"OK, but how does that solve the horizon problem? And what about the structure problem you mentioned before we went surfing?"

"Inflationary theory says that the different regions of the universe we see used to be close enough to communicate, because the little 'patch' was so incredibly tiny. During inflation, space expanded so rapidly that these close regions were spread out to cover all of the visible universe."

"Got it," said Seb. "That just leaves the structure problem."

Chapter 5 – Galloping Inflation

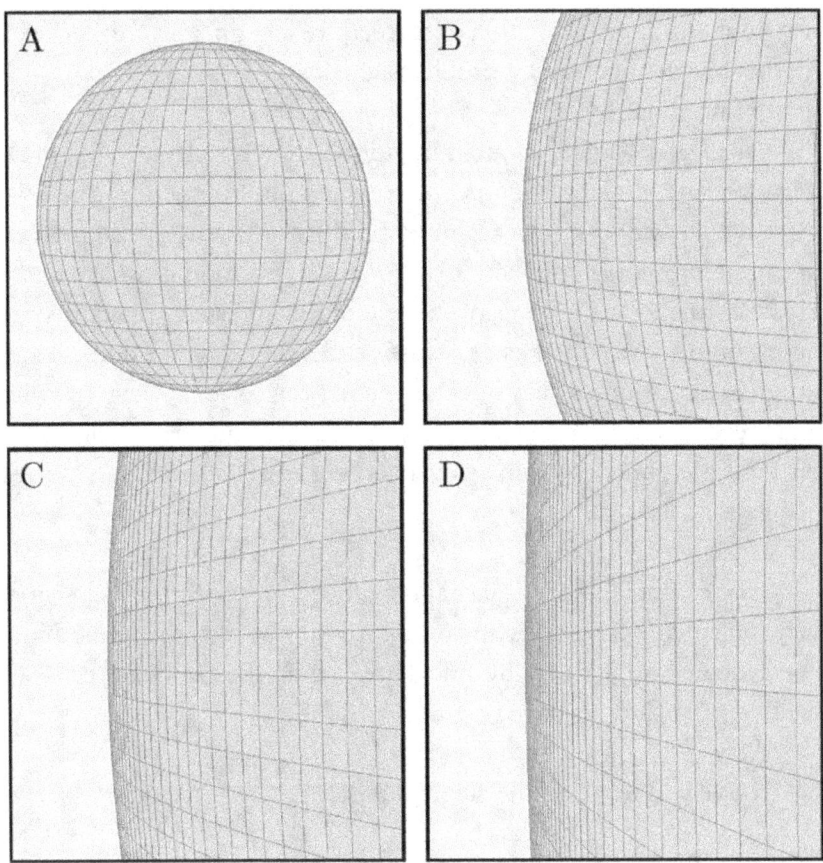

Figure 5.1 Solving the flatness problem
As the sphere gets larger from A to D, its surface becomes progressively flatter.

"Back to our balloon. This time, imagine that it's also very wrinkled. Those wrinkles are the roughness needed in the original materials to make the universe's structure possible. After the balloon has been inflated, every section will be quite smooth and resemble every other section of the balloon. Same goes for the early universe. This, says inflationary theory, is why now, all areas of the universe have essentially the same structure."

"So is inflation the answer? Now we know what happened at the very beginning?" he asked, eyes widening.

"Actually, no. If inflation is correct, we can go a bit further back in time, but the fine-tuning problem remains."

"How come?"

I shrugged. "Think about it. Inflation had to start from something. Brian Greene, a highly regarded theoretical physicist, put it like this: 'from a wild and energetic realm of primordial chaos, there emerged an ultramicroscopic fluctuation of uniform inflation field, and this initiated inflationary expansion.'[2] It lasted a miniscule fraction of a second. But we're still stuck, because as Greene says, 'we don't know how this fluctuation could have happened nor exactly what this wild initial realm was.'[3] In other words, we've pushed the fine-tuning issues to a bit earlier, but we haven't solved all of them. Also, inflation creates problems of its own."

"Like what?"

"For example, it needs to have switched on at just the right time, then remained nearly constant over time to produce the smoothing, then switched off at the right time. However, in the Big Bang scenario, this is very unlikely to have occurred.

"Why?"

"Have you heard of the singularity?"

"Yes, but I'm not sure what it is."

"It's what happens when you run Einstein's equations all the way back to time zero.[4] At that point, you get a singularity of infinitely small space and infinite density, and all the laws of physics break down. It's a point in time without order or rules. Anything can come of it. So we'd expect this point in time to be a universe unthinkably chaotic, like Greene describes, with huge temperature fluctuations from point to point. Clearly in this situation, the strength of inflation wouldn't have been the same in all regions, due to the chaotic quantum fluctuations. In this scenario, either not enough inflation would have happened to smooth things out, or so much would have happened that a very different universe would have developed."[5]

"That sounds like fine-tuning again," said Seb.

I nodded, then glanced sideways at him. "I don't want to depress you, but we also still can't explain where time and space came from."

"Ugh! So what are the latest ideas to deal with fine-tuning then?"

"Some scientists propose a multiverse situation."

"How would that work?"

"Those in favor of this view postulate that instead of the initial parameters of the universe being fine-tuned, they were random. So there are billions and billions of universes, all starting from random parameters, and we live in the one universe where all the parameters are just right. Let's say we're playing dice and we want to roll a six. The only thing we can do is keep rolling until a six shows up. Now imagine rolling twenty-six dice over and over and over until they all came up sixes."

"Twenty-six because that's the number of parameters for the universe?"

"Exactly. But the dice only have six possibilities, whereas the parameters of the universe are very precise numbers—like dice with way more than six numbers on them!"

"How can there be so many universes?"

"Well, if one universe can inflate from primordial chaos, many can just as easily. In fact, according to the inflation theory, there'd be so much energy available that a myriad of universes would be born. So, according to this idea, inflation provides a way to develop many universes, or a multiverse, very quickly."

"Is there a catch?" asked Seb dubiously.

"There is. Science has to be testable. And it's not clear we'll ever be able to 'see' enough of any other part of the multiverse—or at the very least the impact that another universe makes on ours—to confirm the theory."[6]

"Where does that leave science?" he said, sounding a bit exasperated.

"One possibility is that in the future, a deeper understanding of physics will show why the initial parameters of the universe must have had the values we've determined. Unfortunately, in the past forty years, our deeper understanding of physics has actually made the values even more fine-tuned and mysterious."

"It sounds like I need to gain a better understanding of these parameters," Seb said. "Are you up for that?"

By this time, we had reached the hotel and were still sitting in the car, in the underground parking area. "Sure. Let's head to our rooms and get showered and changed first, though."

As we walked up the stairs, I said, "I don't want to leave you with the impression that inflation is a given."

"What do you mean?"

"Well, it hasn't been tested, and recently it's been challenged by observations of gravitational waves."

"Yeah, I saw something online not too long ago that they had finally observed the ripples in space–time that Einstein predicted!"

"Yup, and those observations are causing an issue for the inflation theory. If inflation happened, everything we see originated from a tiny, uniform region of original space. But Einstein's theory says that the more energy the inflation field had, the more space–time would have been shaken by tiny gravitational waves at the beginning of time. So far, observations show no traces of gravitational wave effects from the beginning of the universe, leading us to question whether inflation happened[7]—that is, whether the observable universe was ever smaller than a few millimeters."

Chapter 6

Something Totally Useless

Western Scottish Highlands, summer of 1964

"You're kidding." Peter's voice was flat as he continued peering out of the tent's entrance. The rain was coming down in sheets, and the clouds hung so low that the nearby Highland peaks were invisible.

Behind him, Jody tried, with little success, to stifle her laughter. "I wish I were. Daphne will be absolutely mortified when I tell her! She obviously misread the guidebook."

"Well, she probably should strike travel agent off her list of career possibilities," said Peter, turning toward Jody, his eyebrows raised in amusement. "Even if the downpour eases off this afternoon—and right now, that looks rather unlikely—the trails will be muddy messes. What do you think, Jo? Do we stay and start digging a trench around the tent, or do we head for the first pub between here and Edinburgh?"

"You don't have to ask me twice," she replied, already rolling up her sleeping bag and tucking her copy of *Highlands of Scotland* into the rucksack. "For the life of me, I can't figure out how she thought the book said *lowest* rainfall in Scotland. It's baffling."

"You're the linguist," smiled Peter. "Perhaps Daphne has undiagnosed dyslexia. Now... I hope I remembered to bring extra socks."

Their weekend plans scuttled, the couple drove southeast, swapping seats halfway to make the four-hour drive pleasant rather than tiring. At some point, their conversation turned to a paper Peter had written, "Broken Symmetries, Massless Particles and Gauge Fields," recently published in *Physics Letters*, a journal edited at CERN—Europe's leading nuclear research facility. "Have you had any feedback on it?" asked Jody. Although linguistics was far removed from particle physics, her activism with the Campaign for Nuclear

Disarmament had prompted her to go beyond the average layperson's understanding of that field. And after she and Peter had met at a CND meeting, Jody's interest in theoretical physics had deepened in step with their relationship.

Peter shrugged. "Hardly anything. I think I may need to expand it a bit. I'm convinced the formulae are correct, but I suspect the paper may be a bit too dense, not reader friendly enough. Since we've had to abandon the camping trip, I can spend the weekend writing. I think a second paper, explaining the implications of the theory, may gain a wider audience." His grip on the steering wheel tightened a bit as he began mulling over a structure for the new paper.

Jody suppressed a sigh and looked out the passenger window; she'd been hoping they would make up for the rained-out camping trip by going for a nice long ramble in the hills around Edinburgh tomorrow, having a picnic, taking some photographs and enjoying the unusual late-summer warmth that Scotland—apart from the Highlands—had been experiencing. But she already knew Peter well enough to recognize that bit-between-his-teeth look. Better to leave him to his pencil, paper, tea, and biscuits, and spend the time outdoors with some of her girlfriends.

≈≈≈

Several weeks later, as November drizzle wended its way down the sitting room windows and the summer's warmth was difficult to recall, Peter reached down to pick up the afternoon postal delivery from the mat by the front door. A Swiss stamp and Geneva postmark beckoned, and with a couple of quick tugs of the paper knife, he sliced open the envelope and extracted its contents.

"Dear Dr. Higgs," he read below the CERN letterhead. "Thank you for your recent submission to *Physics Letters*. Regrettably, the editorial board feels that the paper 'Broken Symmetries and the Masses of Gauge Bosons' is not suitable for our rapid publication schedule. We suggest you submit it to a more fitting journal."

Peter sat down heavily in an overstuffed chair, took a poker from the hearth tools, and prodded the fire, moving one of the logs to improve the air flow. As the flames leaped, he put the poker back and shook his head, a slightly rueful expression crossing his face. It sounded to him like he was being politely fobbed off. Well, no matter. He would send the paper off to another journal, this time to one in America.

An hour or so later, after talking on the phone with Jody, who urged him to "add some sales talk" to the paper—"You know us Americans. We like a good sales pitch"—he made a pot of tea, fixed himself some sandwiches, then sat down at his desk and spent the early evening drafting a couple of additional paragraphs for the paper.

In the scant few weeks that it took for his revised manuscript to reach the editorial board of *Physical Review Letters* at the American Physical Society, Peter was told by a colleague with an ear to the ground at CERN that the editors of *Physics Letters* had rejected his second paper because they felt it was "of no obvious relevance to physics." Peter had felt mildly indignant, but arrogance wasn't in his nature. Despite feeling quite certain about what soon came to be known as the Higgs mechanism and the Higgs boson, he'd nonetheless written to one of his PhD students: "This summer, I have found something that is totally useless."

But this "something" proved anything but useless. In the nearly forty-eight years that elapsed between those two groundbreaking papers and the discovery of his eponymous boson, Higgs himself remained relatively unknown outside the circle of particle physicists, although he and "his" particle were drawn into the broader public eye in 2000 when famed professor Stephen Hawking bet one of Higgs's colleagues £100 that the elusive particle would never be found. Yet the search for the "god particle" continued. Higgs eschewed this label, coined by Nobel Prize-winner Leon Lederman, saying, "I'm not a believer in God, but I thought his rather flippant use of the term might be offensive to some people." For better or worse, the catchy phrase stuck.

So why the big song and dance about the Higgs boson? How can an elementary particle with a mass of ~125 giga electron volts and a theoretical average lifetime of only 1.56×10^{-22} seconds (that's 1.56 with twenty-two zeros in front of it) be so important that it featured in an episode of the astronomically popular American sitcom *The Big Bang*

Theory? Put very simply: because without it, nothing would exist (hence the facetious moniker "god particle"). Put a little less simply: because it completes what's known as the Standard Model of particle physics. And this model helps us understand what it takes to build our universe.

≈≈≈

"Did Peter Higgs win a Nobel Prize?" asked Seb, pointing at my iPad. We were back at our hotel to shower and change after a couple of hours of surfing, and he'd asked to take a look at what I'd been writing recently.

"Yes, in 2013, along with François Englert."

Figure 6.1 Large Hadron Collider at CERN

"I'd like to better understand the Standard Model," he said, handing the iPad to me.

"CERN has a short video we could watch if you want." I found the link in my browser history and tapped on it when Seb nodded.

The video began:[1]

> This is one of the most important announcements in the past decades. It confirms, essentially, most of what has been done in physics.

Chapter 6 – Something Totally Useless

> This year's prize is about something very small that makes all the difference. "I'm rather surprised this has happened in my lifetime."

"That's Peter Higgs?" asked Seb, pointing to the screen. I nodded.

> At CERN's foundation in 1954, the world of particle physics was a very different one. Scientists were trying to come to grips with the plethora of particles observed in nature. They lacked an overall framework to explain these basic constituents of matter and the forces that act upon them. This framework would later become known as the Standard Model.

"I don't remember hearing about the Standard Model in high school," said Seb.

I tapped the pause button. "Let's finish watching the video, then if there are things you don't understand, I'll try to explain them." He nodded.

> By the end of the decade, CERN physicists were already providing insights into the weak interaction, a force without which the sun would not shine. The 1960s saw the birth of electroweak theory, which unifies the weak and electromagnetic forces. A vital part of this is a mechanism that accounts for the vastly different ranges of these forces, as well as for particle masses. These were beautiful concepts, but they needed experimental evidence to back them up.

> Particle physicists embarked on a global search for the carriers of the weak force, W and Z bosons, whose existence would prove the theorists right. A major breakthrough came in 1973, with a discovery ... weak neutral currents, telltale signs of the existence of Z bosons. This groundbreaking result brought the first evidence for electroweak theory. Our physicists were on the right track,

but direct detection of weak bosons would take another decade.

Seb interrupted, tapping the pause button. "The electromagnetic force is the one that makes magnets work, right?"

"Yes, and it also makes electrical charges attract and repel, like the nucleus of an atom, with its protons holding on to orbiting electrons."

"OK, but what is this weak force? I've never heard of it."

"It's actually a nuclear force that governs some interactions between subatomic particles. For example, it's responsible for the radioactive decay of atoms—like the decay of carbon-14, which is used to determine the age of certain things."

"All right." Seb hit play.

> By 1983, Super Proton Synchrotron experiments had seen W and Z bosons. Their long-awaited discovery led to the Nobel Prize for CERN's Carlo Rubbia and Simone van der Meer.
>
> The next step began when the 27-kilometer Large Electron-Positron collider was switched on in 1989. It was designed to study weak bosons in detail. The researchers soon had their first major result. By measuring the decays of Z bosons, they found that nature has three, and only three, families of matter particles. Everything we see in the universe is made of the lightest family.
>
> During its 11 years of operation, this collider placed electroweak theory on solid experimental ground. The Standard Model was almost complete, but what accounted for the mass of particles? There was one last missing piece of the puzzle to uncover: the Higgs boson.
>
> Its discovery was in sight. With the construction of the Large Hadron Collider, CERN would take its first steps into a new century of discovery.
>
> On the 4th of July 2012, CERN researchers announced the discovery of Higgs bosons.

Chapter 6 – Something Totally Useless

> It was the final evidence the world had been waiting for, hitting headlines around the world, winning the Nobel Prize for Peter Higgs and François Englert, and cementing CERN's pivotal role in the development of the Standard Model. The Higgs boson completes the Standard Model, but many questions about our universe remain unanswered, mysteries that will captivate future generations of scientists and lead them to untold discoveries.

Seb looked puzzled and intrigued. "I really need you to explain this Standard Model they're talking about. And what do they mean by 'many questions about our universe remain unanswered'? I thought the Higgs particle completed the model."

I set aside the iPad. "Science has established that everything in the universe is made from a few basic building blocks called fundamental particles, governed by four fundamental forces. The Standard Model is the theory that classifies all seventeen known elementary particles and describes three of the four known fundamental forces in the universe. The particles can have mass, charge, and spin."

"Hold on. Can you tell me about the different types of particles first? I've heard of some, but not seventeen!"

"Sure. Let's start with what makes up the matter we see every day. You already know that an atom consists of a nucleus and one or more electrons, and that the nucleus is made of neutrons and protons. These, in turn, are made of just two types of quarks, called *up* and *down*. It's that simple: everything you see is made up of these three particles—the two quarks and the electron."

"Like three types of Lego bricks repeated over and over again?"

"Exactly."

"You said they can have mass, charge, and spin. Does the model tell us these values for each particle?"

"No, we've determined them from experiments. We have no idea why they have the values they do! That's one of the unanswered questions they refer to in the video. The elementary particles of matter are called fermions, and these are divided into two basic groups: quarks and leptons. Each group contains six particles, which are related in

59

pairs known as 'generations.' The lightest and most stable particles make up the first generation; the heavier and less stable particles belong to the second and third generations. Heavier particles quickly decay to the next most stable level."

Seb had been keeping track on his fingers. "That accounts for twelve of the elementary particles, plus the Higgs boson, which makes thirteen."

"The other four particles relate to three of the four fundamental forces at work in the universe, which are the electromagnetic force, the strong force, and the weak force. The fourth, gravity, isn't described by the Standard Model."

"One of the unanswered questions?"

"Yup. Each force works over a different range and has a different strength. Gravity is the weakest, but its range is infinite. The electromagnetic force also has an infinite range but is many times stronger than gravity; like I said earlier, it holds the atom together through protons attracting electrons. The weak and strong forces are effective only over very short microscopic ranges. The strong force holds the protons, neutrons, and nucleus together. The weak force is still much stronger than gravity."

"But where do the other particles come in?"

I smiled. "Hold on, I'm getting there. The forces result from the exchange of force-carrier particles, which belong to a group called 'bosons.' Each fundamental force has its own corresponding boson. The electromagnetic force is carried by the photon, which everyone learns about in their high school science class because it's the particle of light. The gluon carries the strong force, and the W and Z bosons, which get mentioned in the CERN video, give rise to the weak force. We don't know whether a particle exists for gravity. If it does, we haven't discovered it. Another lingering unknown."

"So how does the Higgs boson fit in?" Seb's brow crinkled.

"Well, no one could figure out how all these particles got mass. That was Peter Higgs's brilliance. The boson named after him gives particles their mass. But don't ask me to explain that—it's complicated enough that even experts initially didn't understand. The important thing is we need the Higgs boson for all the other particles to have mass—otherwise, we wouldn't exist."

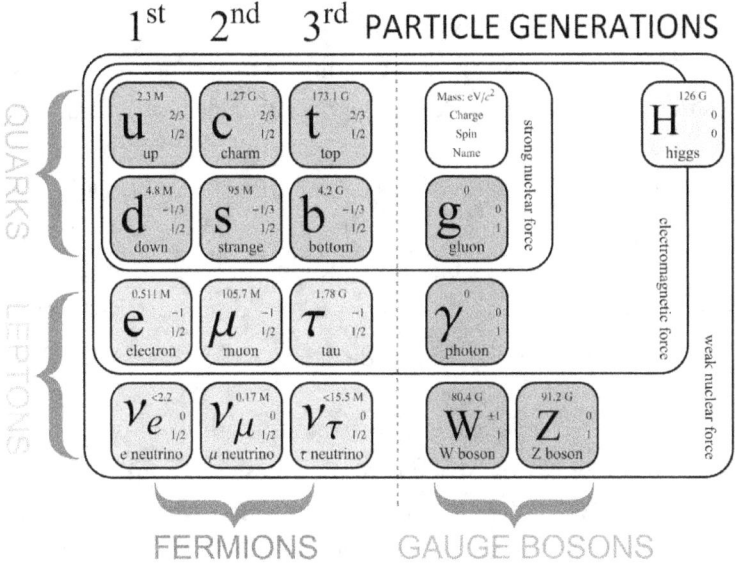

Figure 6.2 The Standard Model of particle physics[c]

"So the Lego set is complete then, except for the couple of things you mentioned?" asked Seb.

"No—we actually still have lots to learn. Why are there three generations, from light and stable to heavier and less stable? Why do all the particles have the particular mass, charge, and spin values they do, and why are those values just right for making our universe exist?"

"The fine-tuning thing again!" he exclaimed.

I nodded. "And why doesn't the Standard Model deal with dark matter and gravity? We simply don't know." I glanced at him. "Have I overloaded you?"

"No, not at all! I'm just processing all of this. Please, carry on. It's fascinating."

"Well, when we study how the things in the universe behave under the forces of nature, we're in for two more surprises. First, the physics that governs the microscopic world of particles—called quantum

---

[c] For a full description of the particles and properties, see Appendix A.

mechanics—is dramatically different from classical physics—in particular Einstein's general relativity—which governs the macroscopic objects in our everyday life. At this point in scientific knowledge, they're incompatible with each other. So we can use quantum mechanics to work all our collider experiments and classical physics to figure out the motion of planets, galaxies, cars, and bicycles, but we can't figure out anything when the microscopic and macroscopic are both relevant, like near the center of a black hole. Second, the physics of objects moving at a normal speed versus the physics of objects such as particles moving at close to the speed of light—what we call relativistic speed—is also dramatically different."

"Yes, I've heard that if things go really fast, time slows down for them, right?"

I nodded. "But let's leave that for a moment until we finish with quantum mechanics versus classical physics. Think about the macroscopic world that we've all known and gotten used to since the day we were born. We intuitively understand how things move and behave. A plane, for instance, is either on the runway or in the air; we know it can't be in both places at the same time. But when it comes to very small particles, our intuition breaks down. These small particles behave as if they were unconstrained by time or space. They can even act as if they're in more than one place at once and as if they affect past events!"

"That sounds nuts."

"Actually, Einstein had a hard time accepting it, but experiments prove that it's true. For example, it's possible for two particles to affect each other instantaneously over a large distance, so large that not even a light signal could travel between them quickly enough to pass on any information. In this way, they behave as if they're connected and right beside each other—like they're 'entangled.' Similarly, a single particle can behave as if it has gone through two different holes in a wall at the same time—as if it split and then came back together!"

Chapter 6 – Something Totally Useless

		Speed	
		Far less than speed of light	Close to speed of light
Size	Macroscopic	**Classical Physics** — classical mechanics and electrodynamics	**Relativistic Mechanics** — special relativity and general relativity - force of gravity
	Microscopic less than 0.000000001 m	**Quantum Mechanics**	**Quantum Field Theory** electroweak - weak and electromagnetic forces QCD - strong force

Standard Model:

u	c	t	γ
d	s	b	g
v_e	v_μ	v_τ	Z
e	μ	τ	W

Figure 6.3 The laws of physics

"The fact that the classical and quantum worlds don't coexist in a single theory is at the root of why science can't explain the beginning of the universe. The microscopic world of quantum mechanics comes together with the macroscopic world of Einstein's theory of gravity, and there doesn't seem to be any crossover. We're stuck with this mystery, although many scientists hope the theories of the macroscopic and microscopic will come together."

"That's frustrating!" exclaimed Seb. "And if it's frustrating for me, it must drive some scientists crazy."

"Maybe. But it also drives them to keep searching, keep experimenting. If you're still OK with all this information, I can tell you about one other key mystery." He gestured for me to go on.

"As we move into the realm of very high speed—where scientists use relativistic mechanics and quantum field theory—various effects arise. The most well known is that time slows down; consequently, particles moving at near the speed of light 'live' longer, meaning they decay more slowly than those not moving so fast. Time actually slows down for them. Macroscopic objects, such as GPS satellites, experience the same effect. Ultimately, at the speed of light, time stops. The particles of light we see that left the early universe thirteen billion years ago, according to our clocks, have actually not experienced even a fraction of a second themselves."

"Why? Can nothing go faster than the speed of light?"

"That's what we think. Light seems to be the speed limit of everything in the universe. Why is it 300,000 kilometers per second? We have no idea, and we don't know why at that speed, time doesn't pass. All we can do is observe that when we assume the speed of light is the limit, we can make the correct calculations that agree with reality. So we're left with very powerful laws and can create amazing technology, but we have little understanding of why these laws are the way they are, and even less understanding of why particles behave, as Einstein put it, in such 'spooky' ways."

"Did he actually say that?" asked Seb.

"Yes. He wrote it in a letter to his close friend Max Born, in German—*spukhafte Fernwirkung*—which is typically translated as 'spooky action at a distance.'"

"I find the whole time-not-passing thing kind of spooky too," said Seb. "I've taken for granted that time is always the same. I mean, it seems to go super-fast when I'm writing an exam, and super-slow when I'm working out at the gym, but I know that's not what you're talking about."

I smiled. "Don't feel bad if you're a bit baffled. Scientists know that time doesn't elapse for something traveling at the speed of light,

Chapter 6 – Something Totally Useless

but even they aren't quite sure what we mean by this, because we're not sure what time is!"

"Seriously?"

"Seriously. If you went to the National Institute of Standards and Technology, in Boulder, Colorado, you could see the atomic clock that standardizes time for the United States. But if you asked the staff there how accurately their clocks measure time, they'd say, 'Our clocks don't measure time. They *define* time with their clicks.'" I reached for the iPad, opened a file, and read from my notes: "One second is the time that elapses during 9,192,631,770 cycles of the radiation produced by the transition between two levels of the cesium 133 atom.[2] So we mark time by the change or movement of something: Earth rotating on its axis, the moon rotating around Earth, the swing of a clock's pendulum. If we want high accuracy, we use an atomic clock."

Seb shook his head. "It's amazing that something I totally take for granted is really so mysterious."

I nodded. "We don't even know whether time is universal. And although we've made progress in understanding why time flows in only one direction, we still don't have a solid explanation. It turns out that the equations of physics work just as well with time running backward, so it's not clear why time's 'arrow' points only to the future."

"Very weird," he said, rubbing his forehead. "How about space? Is that any simpler?"

"Actually, we know even less about space. Our laws of physics don't explain why it exists. We don't know where it came from or what it could have been before. Like we do with time, we measure space, but using a convention that scientists around the world agree on: the definition of a meter. We define a meter as the length of the path traveled by light in a vacuum during about one three-hundred-millionth of a second."[3]

"So scientists are still a long way from getting to grips with time and space?"

"Yep. But hopefully you can now see how incredibly successful the scientific method has been at explaining most of what exists and how it came to be. It's clear that scientific discoveries will continue to provide new knowledge, such as the nature of dark matter."

"Sure," said Seb, "but it also seems like we may never understand the beginning—you know, how the universe started, how time, and space, and the elementary particles, and the forces of nature came to be, and why the particles and forces have their particular fine-tuned values. Maybe we'll never know why the laws of physics are the way they are!"

"I'm inclined to agree. And that's not all…"

"You've got to be kidding!"

I grinned. "Well, it's a vast universe. We've discussed many of the unsolved problems currently being researched, but not all of them. There are other puzzles, such as why today, there is only matter. The Big Bang theory predicts that equal amounts of antimatter and matter were created in the early universe; but if that were the case, we wouldn't exist because matter and antimatter react and destroy each other, generating only energy."

Seb exhaled and said, "I'm starving."

"How about we get some dinner and then, if you're ready, we can go check out the inventor's notes. Maybe the answers to the questions we've encountered are there."

"Sounds great!"

Chapter 7

Master of All the Forces

Safed, Damascus Eyalet, Ottoman Empire, 1569 CE

A slight, striking young woman with diminutive hands held a large and delicately embroidered piece of fabric up to the window and smiled as the early morning sunbeams filtered through its finely woven threads. Cairo had multitudes of weavers and even more embroiderers, but in three decades of living there, she had witnessed only a handful of craftspeople who could produce such exquisite work, such art. Yet after less than a week in the distant town of Safed, she had already seen—and purchased—more superb textiles than she could have found in a year of diligent searching through the Egyptian capital's high-end dealers. This headscarf would be perfect for tomorrow, as she entered for the first time Safed's largest synagogue, where her husband would deliver a speech of thanks to the community for warmly welcoming them.

Breathing in deeply, she savored again the knowledge that they had arrived in their new home and were embarking on a fresh, uncharted part of their life together as a family. Serene by temperament, she nonetheless felt a new level of peace. She had missed her husband tremendously during the past six years, as had their children. Now, although she sensed his time would increasingly be drawn upon by his spiritual followers, Isaac would nonetheless be *home* each evening.

For those six years, her husband had lived in almost complete seclusion on the small island of Jazirat al-Rawda in the Nile. Her father—Isaac's uncle and father-in-law—Mordecai Franses owned the island. So when Isaac, becoming increasingly absorbed by his religious studies, especially the *Zohar*,[d] had decided he wished to live in seclusion to immerse himself in reading, thinking, and writing, the island had been a natural, welcome location. He visited his wife and children once a week, on Shabbat, showing them love and affection but speaking as little as possible.

The arrangement perhaps had seemed strange to some, but she was unperturbed. Her husband, though young, was already a well-respected merchant in Cairo, but his reputation as a student of religion had also grown in recent years. She had rejoiced in the profound fulfillment he experienced from these studies. So when Isaac had told her of his plan, she had listened intently, then had taken his hands, kissed them, and said, "If this is what you are called to do, then you know I will support you entirely. Tell me what you need, what I can do to help you prepare." Having grown up together in the same household since they were small children, they understood each other as relatively few couples did. She had known from early on that he was an extraordinary person marked out by God to bring insights and enlightenment to the world.

No, she had never for a moment regretted becoming the wife of Isaac Luria, increasingly referred to by his followers simply as the Ari (Lion). His reputation for wisdom had preceded them during their migration from Cairo to Jerusalem and then on to Safed. Luria's writings and teachings had met with moderate interest and enthusiasm in the city of his birth. "I scarcely know Jerusalem, and she scarcely knows me; this will change, but not right now," he had said one evening. And so, heeding the advice of his spiritual colleagues as well as his business associates, they had settled instead in Safed, a town at that time blessed in many aspects, including as a center for Jewish mysticism.

---

[d] The foundational text of Kabbalah, written by Rabbi Shimon bar Yochai in the second century CE.

Chapter 7 – Master of All the Forces

Figure 7.1 Title page of the first printed edition of the *Zohar* 1558–1560, Mantua

Isaac's nascent interest in mysticism had been catalyzed when one of his business contacts in Mantua had sent him a copy of the first printed edition of the *Zohar*.

Wishing to know more about what ideas were captivating her husband, she had asked him one afternoon in their shaded garden. Setting down a pitcher of cool water after pouring some for them both, she said, "What can you tell me, Isaac, about these studies that you and the other rabbis call Kabbalah?"

Their oldest son, who in a few months would turn thirteen, walked over from where he'd been examining some bees working their way through a cluster of plume thistles. "Yes, Father, I'd like to know about this Kabbalah too!" Seeing his father smile, the boy sat cross-legged at his feet and looked up eagerly.

Taking a few sips of water, Isaac then cleared his throat and said, "Kabbalah is a mystical tradition, meaning that it seeks to understand intrinsic aspects of God and Creation that are beyond what the intellect can comprehend.[1] Our first patriarch, Abraham, about 4,000 years ago received a revelation called the *Sefer Yetzirah*, the *Book of Creation*. Abraham passed this wisdom down through the generations until seven generations later, the children of Israel received the full Torah at Mount Sinai. The Torah contains a revealed dimension, which we all study. In addition, the Torah has a concealed dimension, which contains secrets relating to God and the Creation, particularly the process of Creation.[2] This concealed dimension is the Kabbalah. The process and order in which the universe was brought into being was further revealed by God through the prophets and sages."

Isaac was about to continue when there was a firm knock at the front door, audible through to the garden. The boy jumped up to answer it and within moments was ushering a tall, modestly dressed young man out to them.

"Ah, Chaim!" said Isaac. "It's good to see you. Is it already time for us to walk to the synagogue?"

"Master, I'm a bit early, but there's a small group of students gathered on the road outside your home, and I can tell they're hoping

Chapter 7 – Master of All the Forces

to speak with you. If we leave now, you can talk to them while we walk."

Isaac laughed. "Good thinking. All right, let us be off."

≈≈≈

Upon his arrival in Safed, word of the Ari had spread as quickly as the lightning that rarely spared their hilltop citadel. He attracted both novices and initiates. To the latter, the Ari imparted great secrets, even the esoteric formulas he used to invoke spirits and conjure healing spells. Students and disciples invariably listened with acute attention, as their master committed almost nothing to writing. "It is for you to decide what you wish to keep for later," he said. "I live and think and speak in the now."

Chaim Vital, one of his students, honored this ideal but knew that neither he nor most others were capable of such a practice. So he took copious notes, not in the Ari's presence but afterward, sometimes writing laboriously late into the night lest he leave something out, and often earning a scolding from his wife for using their precious lamp oil. It had been a blissful, heady time, and Vital had felt his young mind and soul expanding each day.

Not every moment was spent in complicated interpretation and discussion. The Ari taught that joy, particularly through fulfilling the commandments, was essential for gaining esoteric knowledge and experiencing the divine; sadness was the greatest impediment to inspiration in a mystic. Of his master's teachings on this, Vital wrote, "When an individual carries out any precept, be it the study of Torah or prayer, he ought to be joyful and spirited, more than if he had acquired money, or found thousands of gold pieces."

In his every exchange with others, Rabbi Luria sought to be patient, kind, and respectful. Anger was to be avoided in all circumstances, even when another failed to observe a religious obligation, for anger, he said, infected the whole soul and replaced it with wrath and evil. If someone acted disrespectfully toward him, he let it pass. And so careful was he to show his respect that he extended this to all living creatures. "My master," Vital had written one day in his diary, "is careful never to destroy any insect, even the smallest and least significant among them, such as fleas and gnats, bees and the like, even if they annoy him."

Figure 7.2 Rabbi Chaim Vital

For over a year, Vital had met each day with a blend of excitement, wonder, and motivation. The present was bright and the future even more so. Then, on a parched day in July, his teacher, the Ari, had come back from a few days' solitary retreat on Mount Meron and had immediately fallen gravely ill. No one knew for certain what wracked his body, although some of the townspeople murmured about plague. He could keep nothing down, not even water, so they could only sponge his burning body and wipe away the blood that he coughed up more and more frequently. Within a few days, his soul had departed.

Chapter 7 – Master of All the Forces

The mystical community was paralyzed with shock. Rabbi Luria had been only thirty-eight, with a young family and decades ahead of him. Vital was in a fog of grief for days and only ate and drank to appease his wife, who worried he would succumb to whatever mysterious ailment his teacher had contracted on the sacred mountain. On the seventh night, he was visited in a dream by the Arizal, the Ari of Blessed Memory. His lips did not move, but Vital heard his teacher's voice in his mind, calmly saying it was time to rouse himself, that the learning and teaching must continue. "Remember, melancholia is an unpleasant personality trait and even in such circumstances as these should be avoided as much as possible. Immerse yourself in obeying the commandments, studying Kabbalah, and helping others acquire knowledge."

When he woke early the next morning, the weight of his teacher's absence was still there, but he knew what must be done and what would sustain him going forward. He couldn't do it alone, though, much as he wished to. Amassing the Arizal's wisdom to preserve for future generations would require the help of others, so in the coming months, Vital, who had swiftly become regarded as the Arizal's prime student, collected the lecture notes of the other disciples and began to compile these along with his own.

Years later, Vital made his collection of the Arizal's teachings available in manuscript form, titling it *Etz Hayim, The Tree of Life*. In eight volumes, it became the core text synthesizing what is now known as Lurianic Kabbalah. Although not published as printed books until 1782, it had already been widely circulated by that point. From the *Zohar* and other early mystical sources, the Arizal had systematically reconceptualized and expanded upon existing concepts in Kabbalah. Among numerous other elaborations, this resulted in enhanced doctrines about the Creation process.

≈≈≈

At this point, Seb held up a hand. We were sitting at a table, partway through an ample, delicious dinner at a Jewish community and education center adjacent to a resort just a few miles from Legoland.

"Wow," he began. "The Arizal was gifted."

"More than that," I added. "It's said that 'through [him] the spirit of God spoke.'"[3]

"But wait a minute—you're saying that nearly 4,000 years ago, way before the Arizal and even before the giving of the Torah at Sinai, the *Book of Creation* already contained details about the universe's formation?"

I nodded. "That's exactly what I'm saying. The *Sefer Yetzirah* explains the building blocks of the universe. In fact, as we get to know the design manual, most of the mystical information that we're going to explore has its source in the *Sefer Yetzirah* and the writings of the Arizal,[4] predating all of modern physics and even some of classical physics! The Arizal formulated Kabbalah into a comprehensive system that today is known as Lurianic Kabbalah. Before him, Kabbalah had been kept within a close circle; with his disciples, he revolutionized the understanding of Kabbalah and popularized its study."

Seb wrinkled his forehead. "Can you show me how we're able to see the design manual through the mystical tradition, and how this helps us understand the way the universe came to be and why it's the way it is?"

"Sure," I said. "Let's start with the first chapter of Genesis. You might not have read it in a while, but you'll probably remember that it's short, only about 800 words." I reached down to my satchel and pulled out the iPad. "The Genesis account begins with three verses." I turned the iPad toward Seb and pointed to three columns of text on the screen—from left to right, English, Hebrew, and a transcription of the Hebrew into the phonetic alphabet.

Chapter 7 – Master of All the Forces

<u>1</u> In the beginning of **God's** creation of the heavens and the earth.	א בְּרֵאשִׁית בָּרָא אֱלֹהִים אֵת הַשָּׁמַיִם וְאֵת הָאָרֶץ:	1 berêšîṯ bārā **ĕlōhîm**; 'ēṯ haš·šāmayim wə'ēṯ hā'āreṣ.
<u>2</u> Now the earth was astonishingly empty, and darkness was on the face of the deep, and the spirit of **God** was hovering over the face of the water.	ב וְהָאָרֶץ הָיְתָה תֹהוּ וָבֹהוּ וְחֹשֶׁךְ עַל־פְּנֵי תְהוֹם וְרוּחַ אֱלֹהִים מְרַחֶפֶת עַל־פְּנֵי הַמָּיִם:	2 wə·hā·ā·reṣ, hāyəṯāh ṯōhū wāḇōhū, wəḥōšeḵ 'al- pənê ṯəhōwm; wərūaḥ **ĕlōhîm**, məraḥep̄eṯ 'al- pənê hammāyim.
<u>3</u> And **God** said, "Let there be light," and there was light.	ג וַיֹּאמֶר אֱלֹהִים יְהִי־אוֹר וַיְהִי־אוֹר:	3 wayyōmer **ĕlōhîm** yə·hî 'ōwr; wayhî- 'ōwr

Figure 7.3 The first three verses of Genesis

"The first verse of Genesis discusses the creation out of nothing of all the material to make the 'earth,' which here means the physical universe, not just our planet. The Hebrew word for this type of creation is *bara*." I pointed to a highlighted word in the first verse. Seb nodded.

"The second verse describes the initial material of the physical universe, *tohu wabohu*, which has no direct translation into English but basically means 'astonishingly empty.'"

"I don't get it," said Seb. "How can the initial material be emptiness?"

"Well, Kabbalah explains *tohu* as a very thin, chaotic substance[5] that everything was made from.[6] This period of the universe is called the world of *tohu*. Don't worry, I promise it'll become clear shortly."

Seb smiled. "In the meantime, I'm going to order some dessert, if that's OK."

"I'll get whatever you're having," I said. Once a server had stopped to take our dessert requests, I resumed. "The third verse is where we start hearing about the chronology of the universe's development."[7]

"'Let there be light'," Seb read out.

75

"This represents what Kabbalah reveals to be the world of *tikkun*, meaning our world, which is one of rectification. In a bit, you'll see that tikkun is our macroscopic world, subject to the macroscopic laws of physics, where the Big Bang theory applies and works well *after* the first instant."

"One thing about the Genesis story," said Seb, "is that it seems as though a supernatural being is making everything happen. We keep hearing 'God was,' and 'God said.' God's like a magician who keeps making things appear. That doesn't sound like science at all."

"Stepping back from the English translation and going back to the Hebrew helps a lot here too," I said, "because in the Hebrew text, we see that God uses different names to help us comprehend what we might call His different aspects or emanations."

"What does that mean?"

"It's easier to understand if we think about describing ourselves to people who don't know us. That's challenging. We have our personal name, such as Seb or Dan, but that doesn't provide much information. Some of the other names we use are a bit more informative, like father, student, engineer, and so on—the different roles that we have in daily life. You, for instance, are usually in the 'role' of nephew with me. When you're with your friends, though, that's a different role. When you're at your part-time job, you're in the role of employee with your boss and the role of sales rep with your customers. Others understand and relate to us differently depending upon what role we're playing in a given situation. And we behave differently and reveal various sides of ourselves."

"So God uses the same approach to try to make scripture clearer?"

"Exactly." I pointed at the iPad. "Take a look at the first three verses again. See the word in bold? It's the name Elokim.[c] That's the name of God used throughout the first chapter of Genesis."

"What does it mean?"

"It means 'master of all the forces.'"[8] Seb raised his eyebrows inquiringly. "We know this because the root word is *el*, which means power. The second part of the name, *hem/him*, indicates 'them'; in this

---

[c] In order to not use God's name in vain, the spelling of the name *Elokim* has been changed throughout the book. In the proper spelling of *Elokim*, the letter *k* is an *h*.

Chapter 7 – Master of All the Forces

case, 'them' means all the other powers. So *Elokim* means 'the Power over all the powers.'"[9]

"All the powers, meaning...?"

"So not only is God the Creator, but He's also the master over all of the forces of nature in the universe."

"Wait, I think I get it now. So, by using this name, God is indicating that He is appearing to act within nature—through the natural world and its laws?"

"Close. He is of course supernatural, but He chose to accomplish the whole of Creation in a way that looks to us like Him acting through nature."

"Since God created natural laws, He can alter them too, right? So, when God acts outside of natural laws, what name is used then?" Seb asked.

"When God refers to Himself by His essential name, YHVH, He might act outside of nature and time."

"I get it," he said.

"Let's go back to the name Elokim. You know that the twenty-two letters of the Hebrew alphabet are also used as numbers, right?" He nodded. "Then let's add up the letters in the name Elokim." I quickly opened another file and pointed to a simple chart that showed the letters and their numerical equivalents:

100	ק	Kuf	10	י	Yod	1	א	Alef
200	ר	Resh	20	כ,ך	Kaf	2	ב	Bet
300	ש	Shin	30	ל	Lamed	3	ג	Gimel
400	ת	Tav	40	מ,ם	Mem	4	ד	Dalet
			50	נ,ן	Nun	5	ה	Hei
			60	ס	Samech	6	ו	Vav
			70	ע	Ayin	7	ז	Zayin
			80	פ,ף	Pe	8	ח	Chet
			90	צ,ץ	Tzadi	9	ט	Tet

Figure 7.4 The Hebrew alphabet and the numerical values of the letters

"Wait, let me do this," Seb smiled. "It's like deciphering a code!" He took a pen out of his jacket pocket and started scribbling on a paper napkin. It didn't take him long to say, "They add up to eighty-six."[f]

"Right. So now, here's the Hebrew word for nature." After a few taps, I pointed to the screen:

הטבע

Seb scribbled again. "Eighty-six!"

I smiled. "Such correspondences are significant, because they indicate the two words share a meaning. So this is further evidence that—"

"When God uses Elokim, it looks to us as if He's acting through nature and natural laws!"

I nodded.

"Are there other nuances like this in the Hebrew that don't come across in English?"

"There are! Remember I mentioned that at the very beginning, in verse one of Genesis, God describes His action with the word *bara*, which means creation from nothing?" He nodded. "Well, the word bara occurs only once in the description of the creation of the physical universe. In every other place, the making of the universe is described with words that mean making something from something else that already exists."

"Like making furniture from wood, or a cake," Seb pointed to the crumbs on his plate, "with ingredients, like flour, eggs, and sugar?"

"Yep. So if we put these two concepts together—making something from something else and appearing to follow the laws of nature—we get what is actually happening in the Genesis account of the development of the universe, after the first verse. And in fact this is the definition of what science studies: how things come from other things via laws of nature."

"I get it!" Seb said excitedly. "When we're studying the universe using the scientific method, it'll seem like everything came from

---

[f] Elokim in Hebrew: 86 = א+ל+ה+י+ם.

something pre-existing, in a cause-and-effect way, obeying a law of nature. So it can be explained successfully by a scientific theory, like cosmologists do with the Big Bang theory—except for the very first instant."

"Right. So now you can see that God doesn't come across as a magician after all. Genesis, apart from verse one,[10] appears to be a totally natural account based on cause and effect, just as science has discovered."

Seb rubbed his chin. "But why would God do this? Why is it like He's hiding behind natural laws?"

"That's tied in with free will. It's crucial that humans have free will. But let's not wade into those waters right now, because I want to point out another significant correspondence between two words. You said it's like God is hiding. Interestingly, the Hebrew word for 'world' is *olam*. When we add up the letters to get the numerical value for *olam*, that value is the same as the one for the Hebrew word meaning 'concealed, hidden.'"

"All of this is fascinating," said Seb. "And much of this comes from Kabbalah?"

"Yes, in the sense that Kabbalah has revealed some key aspects of the Torah, particularly relating to Creation. But what we've discussed just scratches the surface. There's much more."

Seb sat back. "So before dinner, when we were talking about all the questions that science is still seeking to answer, you mentioned us going to the inventor's notes. Are we getting there?"

I smiled. "Definitely, so bear with me. I'd like you to think of how we make things, from abstract idea through to the real object. Take a house. We start with a flash of inspiration, then develop a picture in our mind, then on a piece of paper or a computer, we set out a scale in feet and inches and draw a blueprint, a design, representing the house. We later use the blueprint to measure the materials and build the actual physical house. According to the Bible, we're made in the image of God,[11] which indicates that our creative process is like His.[12] So God set out a blueprint for creating the finite world and guiding history.[13] That blueprint is the Torah."

"But when I look at a blueprint, it only makes sense if it has a scale—like how many feet on the building are represented by one inch on the blueprint. What's the scale in the Torah if it's a blueprint?" Seb asked, looking puzzled.

"Time. The events described in the Torah are placed on a timeline, and Kabbalah helps us understand the pattern, the underlying methodology, of how the timeline is laid out."

"Go on."

"The mystical tradition reveals that God contracts His infinity through ten channels of divine energy, or life force: the *sefirot* (singular *sefirah*).[14] If we look closely, the Creation is designed around some or all of these sefirot—for example, physical arrangements such as our ten fingers and toes, and more abstract concepts such as the Ten Commandments and the ten plagues of Egypt. These correspond to the ten sefirot."

He looked thoughtful. "Yes, the number ten shows up a lot in the Bible. But we count seven days in a week, not ten. Why is that?"

"Seven is also to do with the sefirot. In the realm of physical time, only seven of those ten channels of divine energy manifest.[15] Think of the six days of Creation followed by the seventh day, the Sabbath, when God rested from creating. We do the same by having the six 'ordinary' days of the week, followed by the seventh day, the Sabbath."[16]

"OK, then if, as you've explained, everything after the first instant will seem to us as though it happened naturally in a cause-and-effect way, doesn't that mean the Genesis account should agree with the scientific account of the timeline for the development of the universe?"

I glanced at the iPad to check the time. "That's what we need to look at next. I don't know about you, though, but I'm tired. Shall we call it a night and maybe tackle that tomorrow?"

Seb had already begun yawning at the word tired. "Yeah, I'm pretty sleepy, now that you mention it. I don't think I'll last for more than another hour, tops."

Chapter 8

Balancing the Scales

I woke in the very early hours and for whatever reason couldn't get back to sleep. So I decided to sit out on the balcony of my hotel room and enjoy the cool, breeze-kissed surroundings before the sun rose and turned them into a sultry oven.

The sky was clear, and numerous stars still managed to make themselves seen despite the light pollution radiating from the city of Carlsbad. Looking up at them, I thought about the career trajectory that for decades had kept me focused on space, alongside studying the biblical texts on Creation.

Since graduating with a degree in engineering physics in 1979, I had worked for a global communications and information company heavily involved in space-related technology. To the general public, the best known of its numerous projects and products are the Canadarm used on the NASA Space Shuttle, and the Canadarm2 and Dextre remote manipulator systems on the International Space Station. The Space Shuttle with its robotics had been used several times to service the Hubble Space Telescope after it went into orbit. This telescope has a direct connection to studies of the universe's age and beginnings.

When Hubble was launched, on April 24, 1990, scientists had ballparked the age of the universe at fifteen to twenty-five billion years. The Hubble Key Project, which ran from 1991 to 2000, provided vital data for significantly narrowing this huge range, and by 2006, additional data enabled a relatively precise estimate: 13.7 billion, give or take a few hundred million years. Since then, it has been further refined to between 13.65 and 13.89 billion years. Hubble also helped bring about new discoveries—in particular relating to the accelerating expansion of the universe, now believed to be caused by as yet undiscovered dark energy.

Figure 8.1 The last Hubble repair mission

While these major technological developments and scientific breakthroughs were taking place, the religious mainstream held to a literal interpretation of Genesis—with God creating the universe and everything in it during six twenty-four-hour days. Among Christians, this was known as Young Earth Creationism, and it remained the predominant view. The majority of orthodox Judaism also retained a strict interpretation of Genesis and thus held that everything was about 6,000 years old.

As a believer in the scientific method and the Bible, I felt sure these two timelines had to be reconcilable. It seemed almost impossible, though; how could you possibly bridge such a gap, get from merely thousands of years to billions? In 2009, I had begun to discover how.

That year, I'd read a book by a twentieth-century rabbi named Aryeh Kaplan titled *The Age of the Universe*. Kaplan had spent the first part of his career, to age thirty-one, as a physicist. At that point, he'd become a rabbi, but the empirical mindset and habits that had made him an award-winning physicist remained engrained, and he applied them with equal rigor in his religious scholarship and writings, publishing nearly fifty books in his unfortunately short lifetime.

Chapter 8 – Balancing the Scales

By his own description, Kaplan was a voracious reader, especially of rabbinic literature. In 1976, he received a photocopy of a microfilm of a manuscript held in the then Lenin State Library—now the National Library of Russia—in Moscow. The manuscript was the *Otzar HaChaim*, by a thirteenth-century rabbi and renowned Kabbalist named Isaac ben Samuel of Acre. Kaplan knew that Isaac ben Samuel's work had largely disappeared after his death in the early fourteenth century. Through a mysterious and fascinating series of events, this manuscript—which seems to be the only complete surviving copy in the world—had gone from the eastern shores of the Mediterranean, to Spain, to St. Petersburg.[1]

Kaplan was intrigued and went to work laboriously deciphering and translating the text. Therein, Isaac ben Samuel described discovering that "the world has existed for a very long time. This is to refute the opinion of those who say that the world has not existed more than 49,000 years."[2] In a startling passage, the medieval sage presents calculations for the age of the universe approaching billions of years:

> I, the insignificant Isaac of Akko, have seen fit to record a great mystery that should be kept very well hidden. One of God's days is one thousand years, as it is written, "For a thousand years in your sight are as a day" (Psalms 90:4). Since one of our years is 365¼ days, a year on high is 365,250 of our years. Two years on high is 730,000 of our years. From this, continue multiplying to 49,000 years, each year consisting of 365¼ days, and each supernal [God] day being one thousand of our years, as it is written, "God alone will prevail on that day" (Isaiah 2:11)

Writing six and a half centuries earlier than Kaplan, he had stopped short of spelling out the full numerical implications of the "mystery" he had discovered. But here was a thirteenth-century Kabbalist articulating a magnitude of age for the universe that scientists had not reached until the early twentieth century! This was the serendipitous breakthrough that opened the door to so much else for me. I went to work, delving into a wide range of ancient sources and consulting with current experts.

The Torah timeline is very clear. First came the six days of Creation. On the sixth day, Adam was made.[3] The end of the sixth day marked the start of the first year since Creation, and since then, 5,781 years have elapsed as of 2021. After Day 6, this is relatively easy to explain using the Torah and other scriptural sources. We are given the number of years of each generation after Adam, and scriptural sources agree that these years are the same length as our years. When we add all the years and correlate them with our secular history, we find that 5,781 years[4] have elapsed since the end of Day 6. There is more than one way to calculate this, but all calculations come to approximately 6,000 years.[5]

What about the first six days? There had to be a time-scale factor so that what the Torah contained could be squared with how old things seem to be when we measure them as we do scientifically. This was where Isaac ben Samuel's work, followed by Kaplan's, gave me a foothold. Through a process that I've described in *The Biblical Clock*,[6] I figured out that one Creation day is equivalent to approximately 2.5 billion of our years—or more exactly, to $1{,}000 \times 7{,}000 \times 365.25 = 2.55675$ billion years. The 365.25 is simply the number of days in a year. The factors of 1,000 and 7,000 can be derived from sources that predate any modern science, cosmology, or estimates of the age of the universe.[7]

Based on the scale factor, the age of the universe as measured by scientists is 13.74 billion years. The same scale factor can be used to elucidate the events and timeline for the formation of the universe, as well as the appearance of life on Earth and the prehistory of humans.[8]

I was mentally reviewing this while watching the Pacific Ocean breakers become increasingly visible as dawn approached. Despite not having slept all that much, I felt refreshed and ready to resume the discussion Seb and I had been having the previous night.

When Seb woke, I said. "Want to get some breakfast and figure out our plans for the day?"

"Sure," he said. "Are you still up for continuing last night's discussion too?"

Chapter 8 – Balancing the Scales

"Absolutely."

It didn't take us long to find a place for breakfast. As we ate, I told him about figuring out the time-scale factor.

"So the biblical and scientific timelines can be harmonized!" Seb said, wide-eyed.

"Yup." I tapped on my iPad, then swiveled it around so he could look at a table I'd created a few weeks earlier. "There are the six days of Creation, followed by 6,000 years of history, which will then be followed by the seventh millennium.[9]

Biblical Calendar – Creation time						Biblical Calendar – Human history						7th millennium
Day 1	Day 2	Day 3	Day 4	Day 5	Day 6	1000	2000	3000	4000	5000	6000	7000
Corresponds to 13.74 billion years						Corresponds to 3760 BCE until 2240 CE						

Figure 8.2 The biblical and scientific chronologies

"Based on the scale factor, the six days of Creation correspond to about 13.74 billion years and depict the events and timeline for the formation of the universe, as well as the appearance of life on Earth and the prehistory of humans. The next 6,000 years match the Gregorian calendar from 3760 BCE to 2240 CE. Scripture designates the Messianic Era—also called the End of Days—as starting on or before the biblical year 6000 (2240 CE) and transitioning to the seventh millennium."[10]

"I need to ask a question," said Seb.

"Fire away."

"What do you mean the six days of Creation correspond to about 13.74 billion years? Did 13.74 billion years elapse or not?" Seb sat back, arms crossed and one eyebrow raised.

"One way to view it is that yes, 13.74 billion years elapsed. When you do that, the dates from science match the Bible after the six Creation days are scaled."

"Wow. But what do you mean by 'one way to view it'?"

"Well, it turns out that our sources by and large say the six Creation days were actually six of our days, not billions of years!"

"But how can that be? When scientists measure things, like the age of the sun, they get way more than 6,000 years, so what's that about? Is science wrong?"

"A perfectly logical question," I said. "The Torah explains that things were made ready to use. So when God made Adam, he didn't make a little fetus, he made a twenty-year-old man. When he made the trees in the garden of Eden, he didn't make seeds, he made trees, and if Adam had cut one down and counted its rings, it may have had a hundred rings, leading to the conclusion that it was that many years old—yet it was just a few hours old."

"If scientists aren't measuring the actual age, then what are they measuring?" asked Seb.

"They're measuring some parameter that would take a certain amount of time to appear if it did so naturally. Let's use the sun as an example."

"Speaking of the sun, shall we maybe head to the beach, so we'll catch a breeze and be able to walk in the water? It's already getting hot," he observed, slipping on his sunglasses.

"Sure." We settled the café tab, then headed on foot for the nearest road to the beach.

"So," I continued, "say we're scientists looking at the sun, assuming everything came about naturally. We use satellites to measure exactly how much hydrogen and helium are in it today. We know—assuming that everything came about naturally—how much hydrogen and helium were present at the very beginning of the universe. We also know about the nuclear reaction that converts hydrogen into helium, which is what's going on in the sun, so we can tell how long it takes for a certain amount of hydrogen to change into helium. This means that if we know how much hydrogen there was at the beginning, how much hydrogen there is now, and how long it takes to convert that much hydrogen into helium, we can calculate that to have today's ratios of hydrogen and helium, the sun had to burn for about 4.6 billion years."[11]

"Right," said Seb. "So what's really going on?"

"Scientists aren't actually measuring the real age, they're measuring a parameter that would take a certain amount of time to appear—like the amount of hydrogen and helium in the sun. And if you make that assumption, the result is correct; the parameter *would* take that long to appear. But the Torah says that when God made the

Chapter 8 – Balancing the Scales

sun, he made it ready to use as it is today, with the amount of helium and hydrogen we see. Basically, God made it old.[12] So if we assume the sun came about naturally, and we measure the hydrogen and helium in it, we'll conclude that billions of years had to pass to get to today's amounts, but the Torah says it was made ready to use only 6,000 years ago."

"Where in the Torah do we learn this?" Seb asked.

"It's right in the first chapter of Genesis, verse ten," I smiled. "Just a few words, but they convey so much. Shall I tell you a brief story related to them?"

"Sure!" Seb said. "I'm actually still a bit sore from surfing yesterday, so I was going to suggest waiting until the afternoon to do some more. That gives us plenty of time." He pointed at some people in brightly colored swimsuits, paddling out toward the higher waves. "We could sit here for a while and maybe get some ideas from watching other surfers while we talk."

I nodded agreement. "The story involves Rashi. You'll know who he is from school."

"Of course! Rabbi Shlomo Yitzchaki. His commentaries on the Torah and the Talmud have been in every edition of those books since… I don't know when, but a very long time!" We sat cross-legged on the sand beneath a large palm tree, well shaded. Then I took a long drink from my water bottle, looked out at the waves, and began.

≈≈≈

France, circa 1060 CE

Rashi was a phenomenally wise scholar who by age twenty-five had already gained a reputation as a skilled teacher. He had the tremendous gift of being able to apprehend a scriptural text's message and then distill this into simple language that even children could understand but scholars also found illuminating. People far older than him, long steeped in studies, would exclaim, "When Rashi expounds upon scripture, it becomes as deep as the sea but as clear as a mountain stream!"

Figure 8.3 Rashi

Despite his inspiring talents, he was a very modest and humble person who shied away from such praise. But he did feel both a responsibility and a desire to use his God-given abilities to the fullest. In the yeshivas of his hometown and the surrounding areas, he noticed a recurring struggle to make the meaning of texts accessible without distorting the truths they contained. Contemplating these predicaments, Rashi decided on a way to put his talent to work, but secretly.

He delighted in the company of children, whose artless, clear-eyed questions and unfettered minds motivated him to make the Torah and Talmud comprehensible for everyone. For many years, Rashi had been writing commentaries on those texts. He thought of them as

Chapter 8 – Balancing the Scales

"scribblings," but to find out whether they might be helpful, he had begun a sort of experiment. Carrying with him little more than some changes of clothing, his writing materials, and copies of the sacred texts, he set about traveling far and wide, visiting the yeshivas in every city, town, and village. Rashi would slip into the classes after everyone else so as not to draw attention; if noticed and welcomed, he would introduce himself simply as "Shlomo." Then, he watched for when students had difficulty understanding, or their teacher struggled to elucidate a passage for them.

Returning to wherever he was staying, Rashi would consider how best to elucidate the problematic lines of scripture. With great care, he would then write out the original passage on a strip of parchment and below that add an explanation in simple words. Returning to the school later that day or the next, he would wait for a time when no one was near where the teacher's books were kept, then he would slip the parchment strip into the correct one, leaving just enough of it showing to be noticed.

Making sure he was there for the next class, Rashi was then able to observe how things unfolded as the teacher discovered the mysterious explanatory slip. Invariably, there would be wide-eyed surprise; sometimes, the teacher all but dropped his book. He would then ask the class whether any of them had put the slip there. The students would, of course, be just as baffled as he.

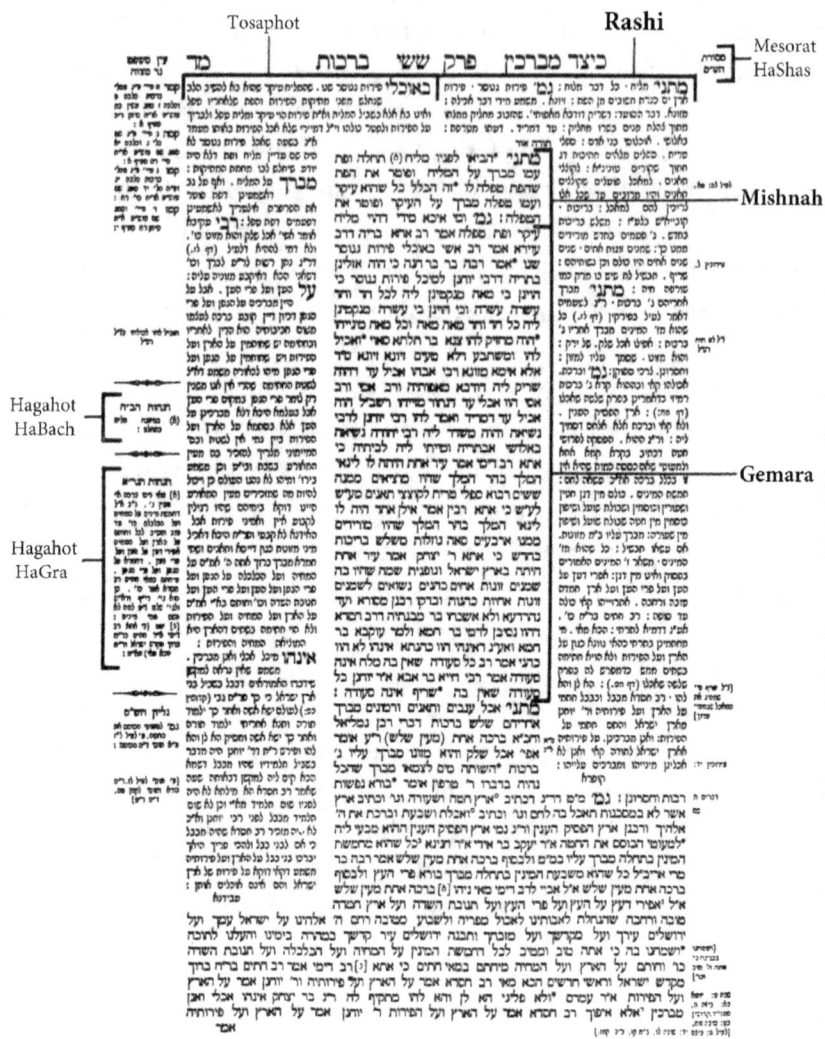

Figure 8.4 A page of the Talmud
showing Rashi's commentary in the right-hand column

But the culmination and most important part of the experiment was when the teacher read out the explanation to his students. Rashi himself always made sure to appear totally absorbed by the book in front of him. But actually, he was gauging the boys' responses. If he saw the dawn of comprehension in their eyes and features, then he was satisfied that his explanation served its purpose well and would become

Chapter 8 – Balancing the Scales

part of his growing collection, which he would at some point organize into a whole that he could leave for coming generations.

Many believed these scriptural annotations were left by God, and to the humble Shlomo's mind, they in effect had been, since it was always God who guided him to his understanding. He was happy to cast his bread upon the water, knowing it would come back as generations upon generations who would be able to learn and understand the Torah and the Talmud with far less toil and much more joy.

≈≈≈

"So," said Seb, "that's how Rashi's commentaries came about?"

"In part, at least," I replied.

"How does this tie in with the world being made ready to use and therefore to look old scientifically?"

"When studying the book of Genesis, Rashi realized the underlying meaning of the phrase 'it was good,' which is used repeatedly in the Creation narrative. Rashi explains that 'it was good' means whatever had been created was ready to use. It had achieved its intended purpose."

"OK," said Seb. "This ties in with where we were last night, talking about how everything after the first instant will seem like it happened naturally, by cause and effect. And the Genesis account of how the universe developed should therefore agree with the scientific account."

I nodded. "To determine when a particular piece of work was completed, and hence when we can expect to see it and measure its age with scientific instruments, we go to the next time the text says 'it was good.' Using Genesis, we can actually determine the times when events happened and generate a timeline for the universe's development during the six Creation days."

"But when I read Genesis, it's really approximate," said Seb. "It just says what day something happened."

"Yes, but the Oral Tradition—remember, this contains the 'design manual' materials—has much more information about timing. Many events in the account can be timed to the second. The least accurate event is the completion of the moon. We only know that it was finished no more than two-thirds of an hour after the sun was finished.[13] I've used the scale factor to compare the times of seventeen events in the Creation days with the scientific times. Take a look at a few well-known events to see how well they match!" I handed Seb the iPad.

Event	Torah calculation[14]	Scientific estimate
Age of the universe	13.743 billion years	13.736 and 13.862 billion years[15]
Age of the Milky Way disk	8.05 to 8.7 billion years	Very approximately 8.3 billion years[16]
Age of the sun	4.79 billion years	4.57–5 billion years[17]
Age of the moon	Less than 70 million years younger than the sun	About 50 million years younger than the sun[18]
Age of the Earth	7.5–8 billion years	4.57 billion years[19]

Figure 8.5 Chronology of the universe

"Keep in mind," I said, "that every Torah source predates any scientific measurement of these events. The only discrepancy relates to the age of the Earth."

"What happens there?"

"Well, in the case of our planet, we're out by billions of years!"

"Really? Why?"

"It could be that I'm doing something wrong, or maybe the current scientifically determined age of the Earth isn't right."

"How can that be?" asked Seb

"Scientifically, we determine the age of most cosmological events from direct measurement, like we do to calculate the age of the sun. But when it comes to Earth, we can't do this, because it's a living planet, constantly destroying and recycling old rocks and creating new rocks through the processes of erosion, plate tectonics, and volcanism. If any of Earth's primordial rocks are left in their original state, they

Chapter 8 – Balancing the Scales

haven't yet been found. So all we can say is that the Earth is at least as old as its oldest rocks, which date to about four billion years ago. However, for very good reasons, scientists have theorized that the Earth formed at the same time as the rest of the solar system, about 4.5 billion years ago."

Seb tapped my arm and pointed out at the waves, drawing my attention to a particularly big one that a few of the surfers were getting ready to ride. We watched as a couple of them handled it beautifully. "I hope we can get some like that this afternoon!" he remarked. "OK, so getting back to the solar system's age… What do they measure to figure out that?"

"Meteorites."

"Fragments of asteroids that fall to Earth?"

I nodded. "These date back to about 4.5 billion years ago."

"So they take the age of the meteorites found on Earth to be the same as the age of Earth?"[20]

"Yes. But there's another possibility. Cosmologists have discovered planets that were ejected from star systems, leading some scientists to postulate that the Earth itself was ejected by an earlier solar system. We know that nucleosynthesis in stars created enough elements to form Earth-like planets approximately eight billion years ago. So the Earth could have formed much earlier somewhere in the disk of our galaxy, then been trapped by the gravity of the solar system's material as the system formed 4.5 billion years ago. In that scenario, asteroids dating back to the formation of the solar system fell to Earth over time, producing the meteorites that we now date."[21]

"How does that compare with the biblical account?" Seb was looking at me intently.

"Remarkably well! And, although it contradicts the currently accepted theory that the Earth formed with the solar system, it doesn't contradict the scientific measurement of the age of the Earth's rocks, and it's consistent with the Bible. Genesis tells us the Earth was formed sometime in the morning of the third day, so one Creation day earlier than the sun and moon, corresponding to about 7.5 to 8 billion years ago."

"That's intriguing," said Seb. "Basically, then, the Torah and science agree that the universe's development seems to have occurred naturally and can be explained by natural laws and theories." I nodded, so he continued. "And with this scale factor, the biblical and scientific timelines can be reconciled as to when major developments took place, like the formation of the sun and moon."

"Yes. That said, each side thinks their timeline is right."[22]

Seb then crossed his arms and looked at me with mock severity. "So Dan, are we *finally* going to get to the first instant?"

I laughed. "We are."

"Let's have it then!"

Chapter 9

Broken Vessels

Kronstadt, Western Russia, 1843

Rabbi Menachem Mendel, the Tzemach Tzedek,[1] beamed as he stood at the front of the room, watching and listening while the two dozen or so soldiers chatted, laughed, and jokingly jostled one another in greeting. At their barracks in the fortified town of Kronstadt, the men had learned that the Tzemach Tzedek was visiting nearby St. Petersburg, so they had insisted that their superiors allow him to meet with them. Not only had this been arranged, but he was going to deliver a Chassidic discourse especially for them.[g]

The soldiers, in their late teens, had been conscripted as children. In the ordinary course of things, they would already be advanced in their Torah studies. But these men's lives had not been ordinary. They were cantonists.

It was 1843, and for sixteen years, Czar Nicholas I's conscription decree had been in effect for Jewish communities throughout Russia. For every 1,000 people, each community had to provide ten boys, aged twelve to twenty-five, to serve in the Russian army. Boys under eighteen were sent to preparatory schools, called "cantons," until they were old enough to enter the military, where they had to serve for twenty-five years. The scarcely veiled primary intent was not to increase the army but to forcibly assimilate and convert. The government employed "catchers" to gather the conscripts wherever Jewish families lived.

As none of the families would willingly surrender their children, those with the means would bribe the catchers not to choose their

---

[g] This event actually occurred, but the discourse presented here, although based on a chapter in one of Rabbi Menachem Mendel's books, is not the one that was actually delivered, the content of which is unknown to the author.

boys, leaving the poorer to have their children literally torn from their arms. In contravention of the decree, children as young as eight were taken; most would never see their families again.

In 1827, the same year of the edict, Rabbi Mendel had been called upon to become the leader, or Rebbe, of the Chabad–Lubavitch movement. With this role came tremendous responsibilities and endless demands. Yet, in addition to all these, he initiated plans to free the captured children by ransoming them from the authorities running the cantons. Forming a secret society called *Techiat Hameitim*, which meant "revivers of the dead," he and others raised huge sums. False death certificates were then issued for the ransomed children, who would be taken into hiding far away from their hometowns for many years, to be nurtured and educated before, if they were lucky, safe reunification with their families.

The Rebbe did this at great personal risk. Undermining the conscription program in any way was regarded as an act of treason, punishable by death. But in the Rebbe's mind there was no question of ceasing their work.

He let the soldiers mingle and joke a little longer than a teacher might have with a regular group, knowing that when it came time for them to settle down and study, discipline wouldn't be an issue. Not only were they eager to learn about their religion and culture, devouring every morsel of information he laid before them, but sadly, all had been terrorized into obedience; every single man bore multiple scars on his torso and limbs from repeated beatings, burnings, and other torture. The Rebbe's eyes clouded over momentarily in sadness, then cleared as he clapped his hands, smiled, and said, "All right. Please sit."

Immediately, the men arranged themselves at their chairs, like iron filings pulled to a strong magnet, and silence fell over the group.

The Rebbe smiled again. "Today," he began, "we're going to explore a topic that one of you raised earlier, during our meeting. It is a very big subject indeed and not something for light conversation." He paused and spread his arms wide. "For what is bigger than the universe and how God brought it into existence?"

Chapter 9 – Broken Vessels

The men remained hushed and intent. The Rebbe then clasped his hands together and continued. "Earlier, we talked very generally about the book of Genesis, and I read out to you the first chapter. Some of you started wondering what the beginning of the universe was like—the very beginning, when God brought something into existence. One of you said, 'What did it look like when God made something from nothing? I've tried picturing the very beginning, but it's perplexing.'"

The Rebbe smiled gently. "Even the Arizal of blessed memory did not believe it is truly possible for us to fully grasp the very beginning or even to imagine what it looked like. Doing this is beyond the capacity of our human intellect. But it's natural for us to try. The order through which the world was brought into being was revealed to us by God through our prophets and the holy sages of blessed memory, as was written down by the Arizal, through whom the spirit of God spoke.[2] I will tell you how it is recorded in *Etz Hayim, The Tree of Life*, a book in which his greatest follower recorded the Arizal's teachings." Stepping over to his desk, the Rebbe picked up a large, beautifully bound volume and held it up for the men to see.

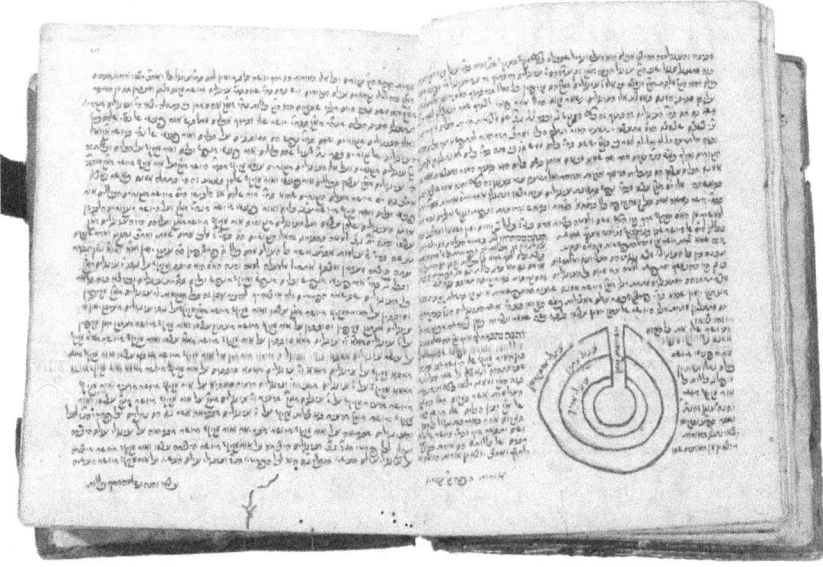

Figure 9.1 Manuscript of *Etz Hayim*, c. 1770 opened to the page on the Creation process

He then continued. "Before the beginning, there was only God and His pure light everywhere, endlessly. We use the idea of light as an analogy. Picture the sun and its rays, then imagine, by analogy, that God the Creator is the sun[h] and He emanates light. His 'light' is referred to as the Ohr Ein Sof, which literally means 'the endless light.' So at the beginning of this supernatural event—creation out of nothing—everything existed in a spiritual plane; there was no physical plane. All was unified, in the ultimate sense, with God. There were neither independent beings nor even limited beings. But when the idea arose in His will to create worlds, there became independent and eventually physical beings.[3] How did this come about? God restricted His light to a central point and left vacant an evenly measured place on all sides.[4] This is called the vacuum and the empty space. In this vacuum there remained only a trace of the original light. Thus, within the vacuum, there was room for his emanations and creations."

The Rebbe looked around the room; every single soldier was rapt with attention. "But Rebbe," began one of the bolder soldiers, raising his hand tentatively, "I don't quite understand what you mean by a trace of the original light remained."

"I'm glad you asked, Shmuel, because this is an area where the writings of the Arizal have been misunderstood by some. It's not that God withdrew all His light—that would have meant removing Himself from Creation[5] rather than just concealing His presence. There is, in fact, 'no place devoid of Him, neither in the upper worlds nor in the lower worlds,'[6] meaning the physical universe.

"Thus, what He did was simply move most of the light to the sides[7] so only a trace was left in this small space. This less intense light allowed for finite creations, whereas the full intensity of the Ohr Ein Sof does not."

Shmuel nodded in understanding, and the Rebbe continued.

---

[h] When discussing the Creation process, Kabbalists liken God to the sun, and His emanations to light from the sun. God is the infinite one, the Ein Sof, and his light is called the Ohr Ein Sof (in Hebrew, ohr means light), which emanates in all directions and fills all space. The light has components, analogous to the colors of the rainbow. His light can be contracted and filtered, analogous to sunglasses acting upon sunlight. The light can go into containers and emit a particular component, like when a candle is in a red vase and we see only red light from the outside. See Appendix B: The Light Metaphor.

Chapter 9 – Broken Vessels

"He created this tiny primordial space, or void, but it wouldn't remain small, because He created it as expanding,[8] getting larger and larger every moment. And as Ramban teaches, the Ohr Ein Sof underwent many contractions, until the void eventually contained a very thin substance, called *tohu*, which had 'a power of potency, from which everything else would be made.'[9] So the basic substance from which everything else in the universe would be made was *created*, which means it came to be something from nothing. It was not *made*, which means something from something. This first, *primordial matter*[10] included the basic ingredients that had the potency to proceed from potentiality to physical reality. So," he paused, "from the first step, infinite to finite, everything else proceeded."[11]

One of the soldiers raised his hand, and the Rebbe nodded at him. "Rebbe, so the very first creation was something from nothing, and then after this, God made the rest of the Creation from that initial something?"

"Almost," said the Rebbe. "The Arizal teaches us about this too. Let's examine the words in Genesis."

| <u>1</u> In the beginning of God's creation of the heavens and the earth. | א בְּרֵאשִׁית בָּרָא אֱלֹהִים אֵת הַשָּׁמַיִם וְאֵת הָאָרֶץ: | 1 bərêšîṯ bārā ʼĕlōhîm; ʼēṯ haš·šāmayim wəʼēṯ hāʼāreṣ. |

Figure 9.2 The first verse of Genesis, in English, Hebrew script, and transliterated Hebrew

"You see," said the Rebbe, "in the first verse, where God describes His action at the very beginning, He uses the word *bara*, which means creation out of nothing. This is different from the words He uses to describe making something from something else that already exists. This word bara occurs only once in the creation of the inanimate universe, right in the beginning sentence. Every other action related to making the universe is a word that describes making something from something. Bara also appears twice more in Genesis, in the narrative about the creation of life."

"Rebbe," said another soldier, "I can understand why God had to act in this way at the very beginning, since nothing existed. But why did he do it two more times, when he'd already brought into existence what he needed to create everything else?"

"That's an excellent question, Saul. We don't have time to go deeply into this, but when it comes to life, unlike the rest of the universe, more than material things are required to make it. For example, when Adam was made, bara is used to indicate that he had a divine soul[12] added to his earthly body. But," he held up a hand and smiled gently, "we're digressing. What's important for now is that the bara events serve a bigger purpose. They're God's way of leaving signs for us—clues that He exists."

"Clues?" said Saul.

"Yes. After creating everything, God didn't wish to be apparent in the physical world to most people, because He wanted humans to have free will and to exercise that. It would be too overwhelming for someone to be in God's presence; that person definitely wouldn't have free choice; he would just do the right thing. Think of when a big fire is raging in front of us; we don't put our hand in the fire, because it's so clear that we would be burned, and in that sense, our choice is taken away. In this world, God's presence isn't obvious to most except if one tries to understand the places where bara occurs. These events, like the beginning, have no natural explanations, only supernatural ones[13]."

"Rebbe," said Shem, the soldier who had expressed perplexity at the earlier meeting, "here we are, at only the first verse of the first book of the Tanakh, and yet already, you've taught us so much!" He looked around at his comrades. "We toiled all morning to prepare for your visit, polishing our buttons in your honor. Now, you're polishing our souls, which have been dulled and coarsened by so many years of disconnection from Jewish life.[i] If only," he lowered his voice almost to a whisper, "you could stay—or we could leave."

---

i   The passage in the source reads: "Rebbe! We've been toiling all morning to prepare for your coming, polishing our buttons in your honor. Now it's your turn to work hard: polish our souls, which have been dulled and coarsened by our many years of disconnection from Jewish life."
See www.chabad.org/library/article_cdo/aid/166403/jewish/Sand-and-Water.htm

Chapter 9 – Broken Vessels

≈≈≈

Seb emitted a low whistle of amazement. "Dan, I'd never heard of that decree or the cantonists. How long was the law in effect?"

"Thirty years," I replied. "Fifty thousand children were torn from their families and 'de-Judaized,' the aim being to make them at least Russian and preferably also Christian."

He shook his head. "It sounds quite a bit like what was done in Canada to First Nations children through the residential school system." Seb was solemn. "I knew of the Tzemach Tzedek but had no idea about his involvement in rescuing those children. Fascinating. And terribly sad."

We were still watching the surf and surfers, although neither of us was paying conscious attention anymore, too absorbed in our conversation.

Seb then sat up straight. "I want to do more reading about that period in history, for sure. But right now, I want to stay on the topic of the first instant, since it took us so long to get here!"

"Fair enough," I said, smiling.

"In the story, the Tzemach Tzedek mentions that Ramban uses *tohu* to describe what was left in the void, after the out of nothing step. Where is that in Genesis?"

"Right in the next line, in Genesis 1:2, where it says 'and the earth'—meaning the universe—'was tohu vabohu.'"

2 Now the earth was astonishingly empty, and darkness was on the face of the deep, and the spirit of **God** was hovering over the face of the water.	ב וְהָאָ֗רֶץ הָיְתָ֥ה תֹ֙הוּ֙ וָבֹ֔הוּ וְחֹ֖שֶׁךְ עַל־פְּנֵ֣י תְה֑וֹם וְר֣וּחַ אֱלֹהִ֔ים מְרַחֶ֖פֶת עַל־פְּנֵ֥י הַמָּֽיִם׃	2 wə·hā'ā·reṣ, hāyətāh tōhū wāḇōhū, wəḥōšek̲ 'al- pənê təhōwm; wərūaḥ 'ĕlōhîm, mərahep̲et 'al- pənê hammāyim.

Figure 9.3 The second verse of Genesis

"You see, for the physical world to eventually appear, the divine light had to go through a process that started with the most sublime—the Ohr Ein Sof—and descended through many contractions, concealments, and steps until finally, the physical world could exist. This process is called tzimzum."

Seb squinted. "Why did the light need to go through so many contractions?"

"Well, imagine the original divine light as a big waterfall. We want to capture some of its water, but can't do this with a cup; the strength of the waterfall will simply knock the cup out of our hands or break it. So we capture it in a big barrel; think of this as a contraction. Then we use a large container to capture water from the barrel; that's another contraction. Then we can use a cup to scoop water from the large container. Similarly, divine light after many contractions can enter a finite physical object."

"I get it now," said Seb, nodding. "But how did the very first step happen?"

"Great question. The analogy is good for most steps, but not the first one, because it was from nothing. The first contraction was a 'leap' from the infinite divine to something non-divine and finite. Only after that leap can the rest of the steps be explained by the analogy, just successive gradual steps."[14]

Seb's brow was wrinkled, but he didn't say anything, so I continued. "First, God had to withdraw His Infinite Light and leave a smaller light so that independent things could exist. Next, God had to put that smaller light into vessels,[15] specifically the vessels of tohu."[16]

"What does tohu mean?" Seb interjected.

"It doesn't have a proper translation into English, but the words 'chaos' and 'formless' are often used. So because these vessels,[17] in the world of tohu,[18] were too weak to contain the light, they broke.[19] As a result, every fragment of the broken vessels had a spark of the Godly light attached to it, and these bits then metamorphosed, in many steps and stages, ultimately producing the physical constituents of matter, the components that make up all of the world."[20,21] I stopped when I saw that Seb looked somewhat alarmed. "What's wrong?"

"How could the vessels have broken? Were they flawed?"

"No, there was no flaw in the creative process, and the shattering of the vessels wasn't a coincidence; it was a destruction for the purpose of building. The shattering of the vessels served a very specific and

Chapter 9 – Broken Vessels

important purpose: to partition the light into distinct qualities and attributes, introducing diversity and multiplicity into Creation. In particular, the constituents of matter—the elementary particles—were created."[22]

"What happened next?" asked Seb.

"God continued the process by putting the now broken fragments with sparks of His light into new vessels. These were vessels of tikkun, which translates literally as 'rectification.' They were stronger,[23] so they could contain the light in the sparks and didn't break. The world we know around us—trees, birds, ocean—are all part of the world of tikkun."

"How did the world of tikkun come about?"

"Well for that, we need Genesis 1:3, where it says: 'and God said, "Let there be light," and there was light.'"

| 3 And **God** said, "Let there be light," and there was light. | ג וַיֹּאמֶר אֱלֹהִים יְהִי־אוֹר וַיְהִי־אוֹר׃ | 3 wayyōmer **ĕlōhîm** yə·hî 'ōwr; wayhî-'ōwr |

Figure 9.4 The third verse of Genesis

"This means it came about by speech, by God's divine word—specifically by combinations of the twenty-two Hebrew letters.[j] As we've talked about, each letter has a shape, sound, and number.[24] So they can be used to make the myriad of things we observe in the macroscopic world."

"Wow, we finally got there!" Seb exhaled. "But I need a summary."

I laughed. "Very understandable. OK, so now you can see that the Torah says the beginning occurred outside of what science can understand. The first act of Creation was bara, something from nothing. Science doesn't deal with nothing.[25] By definition, natural laws are statements that describe or predict a range of phenomena as they appear in nature. These laws all deal with things that already exist. No law of nature speaks of something coming from nothing. However,

---

[j] See Appendix C for a list of the letters, their shapes, and their numerical values.

the entire Genesis account is in the name of Elokim, so it appears to be governed by the laws of nature. Also, the Torah clearly stipulates what came from that very beginning process."

"This is a lot to take in," said Seb. "Can you unpack it a bit more?"

I nodded. "That's why we needed to talk through so many other things first. Let's go back to what the Tzemach Tzedek in my story taught the cantonist boys. First, there was only God. All was unified—in the ultimate sense—with God. There weren't independent beings or even limited beings. But when the idea arose in His will to create worlds, He began His work of Creation[26] by creating all existences in their most sublime and spiritual form. He then caused them to gradually metamorphose, in many steps and stages, ultimately producing the physical world, which is the most tangible embodiment of these results.[27, 28] The contents of the physical world, as well as its defining parameters—space and time—are the end-of-the-line products of this multistage process."[29]

"Right," said Seb. "So, the whole first instant is described in the first three verses of Genesis, which are really short. But the mystical tradition elucidates them for us." He rubbed his forehead. "Your diagrams usually help me visualize the relationships between ideas. You don't happen to have…?"

I reached into my satchel. "Yup. iPad to the rescue again." I smiled and with a few taps, opened up a diagram on the screen.

Chapter 9 – Broken Vessels

Order of Coming into Being	**First Spiritual**	God is "first." His existence is unchanged.				**First Order of Time**
		One Glance			All of history from the beginning to the end	
		Divided into 6 attributes			The single glance is divided	Process or sequence of separate units is one-way
		Day 1			Each attribute is brought down in sequence as days and millennia	
			Day 2	1,000 years		
			...	1,000 years		
		Creation of the Void				
	Physical					**Space**
		Billions of years for the development of the universe	Thousands of years of history		Measured only by movement of physical entities	**Time**

Figure 9.5 The first instant, modified for the void being nonphysical

"But I have to provide a disclaimer first."

"Oh no, just when we're *finally* getting there!" Seb protested.

"I know, I know. Don't worry, we're almost there. The whole explanation about contractions of the Ohr Ein Sof and so on describes the creation steps in the spiritual strata, not the physical world. It's tempting to read it as a physical occurrence, but we're not given the actual physical description, only the spiritual. So, what happened physically in the first instant isn't known. However, as we've seen, the same sources say, 'God began His work of creation by creating all existences in their most sublime and spiritual form. He then proceeded to cause them to evolve and metamorphose, in many steps ... ultimately producing the physical world.'[30] So it's possible that the creation of the physical world reflected that deeper spiritual origin. I've constructed the diagram in this vein—simply assuming that the physical steps reflected the spiritual steps—but that's just me, not the sources."

I pointed to the top of the diagram. "First, God decided on the 'order of time,'[31] the order in which things would be done. He foresaw in 'one glance'[32] all of existence: the six days of Creation and its long list of events, the Sabbath, the 6,000 years of history toward the culmination of the Divine Plan, the Messianic Era—which is to occur by the year 6000—and the world to come, which is the seventh millennium and beyond.[33] God saw all of time in this one glance, all planned around the numbers six and seven. But this wasn't physical time as we know it; this was an order or list of events not yet scheduled."

Biblical Calendar – Creation time						Biblical Calendar – Human history						7th millennium
Day 1	Day 2	Day 3	Day 4	Day 5	Day 6	1000	2000	3000	4000	5000	6000	7000
Corresponds to 13.74 billion years						Corresponds to 3760 BCE until 2240 CE						

Figure 9.6 The biblical and scientific chronologies

"When you say the year 6000, do you mean the biblical year counted from Creation?" Seb said.

"Yes. The 6,000 years of history after the sixth day of Creation correspond to the years 3760 BCE until 2240 CE, and the year 6000 is the year 2240 CE."

"OK. But I'm a bit confused by 'order of time' versus time."

"Then let's imagine we've decided to build a tree house. We begin by making a list of what needs to be done, often in chronological order but not yet according to a schedule. For example, we might write down:

- Select the tree
- Draw up the plan
- Get the materials
- Build

"Next, if we're good builders, we make a schedule of when each of these tasks will be done—which days, at what times—then we proceed."

Chapter 9 – Broken Vessels

"Got it," said Seb.

"God did precisely this. First, he decided on the 'order of time'[34]—so the order in which things would be done, like us with our list of tasks for building the tree house. But physical time didn't exist yet, because physical space had to be the first creation devolving from the 'void.' Once physical space existed, physical time could exist." I pointed at the bottom of the diagram.

"Why was that?"

"Because according to scripture, physical time is nothing more than measured motion—the number of swings of a pendulum, or the revolutions of the Earth, or how many times electrons of atoms inside an atomic clock change energy level. So physical time isn't possible until space and things that move or change in that space exist. Only then can the spiritual order of time be converted to a 'schedule,' to physical time."

"All right, that's making sense to me. It's like what science has discovered, that time is just counting the motion or change of something," said Seb. "But we've talked about the universe being fine-tuned. How does this first step create a fine-tuned universe?"

"Again, great question. Well, it's now clear that what Genesis describes is a process in which everything is planned and fine-tuned for existence. So in a process where something (namely, us and our universe) comes from nothing, it comes in a fine-tuned way that makes it possible for the something—us—to exist. The Creator, like the architect of a house, sets specific parameters so that the whole thing works. When we have a house built, it's fine-tuned for our existence, designed for us: the doors are the right size, and so are the toilets, the kitchen, and so on. How these come together to make up the house isn't a random accident."

Seb was nodding. "The vessels of tohu and the vessels of tikkun were also new to me. How do they tie in with what science observes?"

"When the tohu vessels broke, matter devolved from fragments of the broken vessels, with only a spark of light attached to each.[35] So those elementary particles are unconstrained by time and space."

"Why?"

107

"Because the vessels are what constrain creations to time and space. And elementary particles are standalone—they're not in vessels, they come from broken vessels, so they aren't constrained. This means they behave in the 'weird' ways we talked about when reviewing the microscopic world, where quantum mechanics applies rather than classical mechanics.[36] We saw that the particles violated common sense by not being in just one place at one time. Now we see that they're not bound by space, so they can be in many places, and they're not bound by time, so they can appear to violate the chronology of time, like the spookiness Einstein was talking about. We've also seen that all microscopic constituents are quantized—because they're vessel fragments carrying the sparks of light. They're individual specific 'letters' that aren't bound by time and space, by the vessels.[37]

"While we're on the subject, it's interesting to note that tohu, which approximates in English to 'chaos' and 'formless,' is an almost identical description to the one used by science to describe the original quantum vacuum that existed right after the beginning. For example, a famous science writer, Brian Greene, describes the beginning as 'a wild and energetic realm of primordial chaos.'"[38]

Seb clasped his hands and looked out intently at the waves. "So when these elementary particles are made into bigger objects and put into what derives from the tikkun vessels, which makes up the macroscopic world, then they become constrained by time and space, and everything makes sense to us according to classical physics?"[39]

"That's right, and everything is chronological, just like physics explains for the macroscopic world. The sources also go further, explaining that anything that's a result of bara is perfect, unchanging,[40] and everlasting.[41] This is why the elementary particles live[42] and keep going forever. Science hasn't been able to explain this. For example, a photon moves at the speed of light forever unless it hits something; an electron spins at a particular rate forever, and so forth. Everything else in this world, if left alone, eventually decays back to the original components. The sources put it this way: the rest of Creation, from rocks to plants and so on, is formed—something from something else—and so it can continue to change and decay until it reverts to

Chapter 9 – Broken Vessels

tohu, which is permanent, what science calls the elementary particles."[43]

"Now that we've reached the world of tikkun, which was formed with the twenty-two letters of the Hebrew alphabet, can you tell me a bit more about that?"

"Sure. See, the physical world devolved by many steps from God's spiritual speech."

He held up both hands. "Whoa, that's confusing! God doesn't speak, does He?"

"Well, let's see. The words used in Torah are exact. When it says God spoke, it means something exact. Yet we know God doesn't speak like us. So how are we to understand what 'God spoke' means? By looking at ourselves as the analogy, specifically our process of speaking. When we speak, the process is to take a thought—something infinite and spiritual, not physical—and condense it into a set of words delivered in chronological order. The words are then real objects in an order and can be written down and further refined into instructions to make actual physical things. This is exactly the process of creation that the Torah describes. God contracted his infinity and eventually produced physical things in a chronological order: the six days of Creation. This gives us a glimpse of what 'God spoke' means—it's the creation process we've been studying."

"OK," said Seb, "you haven't lost me. Let's keep going."

I nodded. "Before God's supernal lights could produce physical actions, they had to enter into the mystical realm of letters, which exist to bring things about in the physical realm. This is what's meant in Psalm 33, verse 6, which says: 'The Heavens were fashioned through God's word.'[44] The letters function as exactly twenty-two different categories of phenomena, no less and no more, which exist in order to give the lights the ability to act.[45] While the letters played a role in Creation, 'the combination of them maintains the world.'"[46]

"OK," said Seb. "So, it sounds like the letters function pretty much the way science says the elementary particles do."

"That's the idea, yes. Through many steps, the physical world devolved from the spiritual speech, and everything is made by combining twenty-two entities, described as letters, into larger objects,

109

words. This implies that on the physical plane, there are twenty-two types of elementary particles from which the macroscopic world is made. Creation exists because of a life force vested in it through transposing the letters and their numerical values.[47] Or, as science would put it, the different combinations of elementary particles with all their special fine-tuned properties make everything exist."

"But doesn't the Standard Model say seventeen elementary particles?"

"Yes, but remember, we said it isn't complete—for example, it doesn't deal with dark matter. We'll get to that. But first, we should return to what the design manual, the Torah, says came right after the first instant—the first fundamental components of the universe, which are physical space and physical time."

"I need to process," said Seb, getting up. "Should we get some exercise first?"

"Sounds like a plan. Let's get our gear, then head out into the waves."

Chapter 10

It's About Time

Seb and I flopped down on our large blankets, still a bit short of breath from clambering out of the surf and carrying our boards back to where we'd set up our stuff under a couple of beach umbrellas. As we toweled off, then started to unpack our lunches, I noticed he seemed preoccupied.

"Anything bothering you?" I asked, offering him a drink.

"Thanks," he replied, holding out his cup. "Not bothering me, exactly. I'm just thinking about time and space. How they seem so simple until you start wondering where they came from. Then…" he rubbed his face, "they don't seem straightforward at all."

"Should we leave it for today?"

"No, I definitely want to get back to the design manual and come to grips with this, so let's continue. Food will help; I'm starving." Sticking his hand in the small cooler we'd borrowed from the hotel, he pulled out a sandwich and demolished half of it in a couple of bites.

I smiled. "From the Torah and its commentators, we've established that the first physical creation was space, 'a very small point,' coming from a spiritual entity who 'left vacant an evenly measured place on all sides.'[1] From this, we can ascertain that space was small but finite."

"How small?"

"Based on contextual uses in other parts of the Torah, small means millimeters.[2] So the Torah is saying there was no beginning state of everything being at infinite density and temperature. Instead, the initial Creation was what science would describe as the situation after inflation. This space was then filled with particles, becoming identical to what science calls the quantum vacuum."

"So, science and the Torah agree about this?" asked Seb.

"Yes. By going back in time, as science does, we reach the correct answer: space was dense and packed with particles, but it was finite.

Remember our earlier analogy, about taking apart the house and concluding that wood was the main component, then looking at where the wood came from?" He nodded. "Well, the Torah stops there and says this very small point was the primordial out of nothing space—in other words, the wood came from a tree. Science, though, continues to go back, necessitating the invention of inflation."

"Which scientists aren't actually certain about, right?"

"Right, but if observations challenge its validity, some other natural theory will be proposed. Presently, the inflation theory proceeds to that first instant, then reaches an unexplainable situation; in the house analogy, this was concluding that the wood was put together from carbon atoms rather than being cut from a tree. In addition, the Torah description makes clear that this space is a continuous substance and not quantized, something that scientists are still debating."

"OK, can we pause here for a sec?" Seb said. "How does the Torah description make this clear?"

"We saw that the description of the primordial space was a very small point, expanding. This describes a continuous thing like a small elastic piece being pulled to become larger. This is in contrast to the Torah's description of the elementary particles, which come from the twenty-two Hebrew letters. Each letter is discrete and unique, so these particles are unique or quantized units.

"Remarkably, the Torah also explains that this space is expanding. It made this assertion thousands of years ago, but scientists only recently discovered it." I read from my iPad:[3]

> At the time that the Holy One, blessed be He, created the world, it went on expanding like two clews of warp …

"A clew is a ball of yarn or thread or cord, and 'warp' refers to the lengthwise threads on a loom. These two warps symbolize heaven (the spiritual realm), earth (the physical universe), and the two primordial materials that make each.[4] Incredibly, the Torah then goes on to say that the expansion was brought to a standstill: 'it went on expanding like two clews of warp, until the Holy One, blessed be He, rebuked it and brought it to a standstill.'"

"Wow," said Seb. "When did that happen?"

Chapter 10 – It's About Time

"The sages teach that matter and energy, once called into existence, were in a state of evolution until God set a limit to their development. When this ending of creation was reached, no new formations emerged, and that was the Sabbath of Creation, the seventh day.[5] According to our timeline, that Sabbath was 5,781 years ago, so the universe stopped expanding then, about 6,000 years ago. As Rambam explained,[6] the text may mean that only the laws of nature became fixed then, or if the literal reading is correct, the actual expansion of space stopped then."

Partway through a second sandwich, Seb paused and asked, "If the expansion stopped, why haven't scientists discovered this?"

"Good question. Scientists measure the universe's expansion by looking at stars and measuring how stretched their light is and how far away they are."[7]

"What do you mean by light being stretched? And how do they measure it?"

"When we see light from a galaxy ten billion light years away, it means that the light waves we're observing had to have been emitted ten billion years ago. As the universe is expanding with time, the space through which the emitted light wave travels to get to Earth is itself stretching, meaning the light waves stretch as well."

"You mean their wavelengths become longer?" Seb asked.

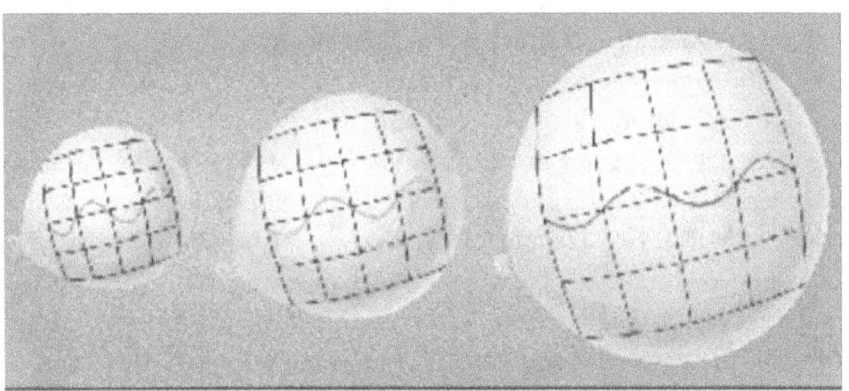

Figure 10.1 The wavelength of light expands as the universe expands.

"That's right. The farther away a source is from Earth, the more time space has had to expand between when the light wave was first emitted and when we observe it. So, if we know how far away many stars are and then measure how stretched their light is, we can determine the expansion rate of the universe over time. The hard part of this process is determining the actual distance to the stars—but that's another topic."[8]

"Getting back to your question of why scientists haven't yet discovered that the universe has recently stopped expanding, well, as we've talked about, when we look at a star that's, say 10,000 light years away, we see the light that left it 10,000 years ago. We don't see it as it is today but as it was 10,000 years earlier. When we measure the expansion of the universe, we use stars that are hundreds of thousands[9] or millions and billions of light years away, and we see the history of expansion over the thirteen billion years of the universe's existence. But we don't see what was happening in the very recent past, specifically in the past 6,000 years."

"Wait, why don't scientists use the light from stars that are 6,000 or less light years away?"

"Because anything that close is gravitationally bound to us and therefore is not affected by the expansion of the universe. Even for the closest objects that are actually receding from us the stretch effect is too small to detect in comparison to their intrinsic movement, which also affects their emitted wavelengths. So, the expansion of the universe is computed from stars much farther away."

"This is pretty amazing," Seb said.

"There's more. The sources also assert that the solar system occupies a special place in the universe, whereas the Big Bang theory assumes that we're not in any special location. This is known as the cosmological principle, which states that the distribution of matter is homogeneous and isotropic when viewed on a large scale. The mystical tradition dating back to the oldest Kabbalistic book, the *Book of Creation*, discusses the special place we occupy. It says the celestial bodies were set 'in the *teli*.'[10] Teli is a very mysterious word, and there's considerable discussion about its interpretation. It's believed to be the axis of the plane of the ecliptic."

"Whoa," Seb said, "what's that?"

Chapter 10 – It's About Time

"The imaginary plane containing the Earth's orbit around the sun." He nodded, so I continued, "The teli is the center around which everything in the universe happens or is arranged—and even perhaps rotates.[11] So the sources are saying we do occupy a special place: everything is arranged around this teli."

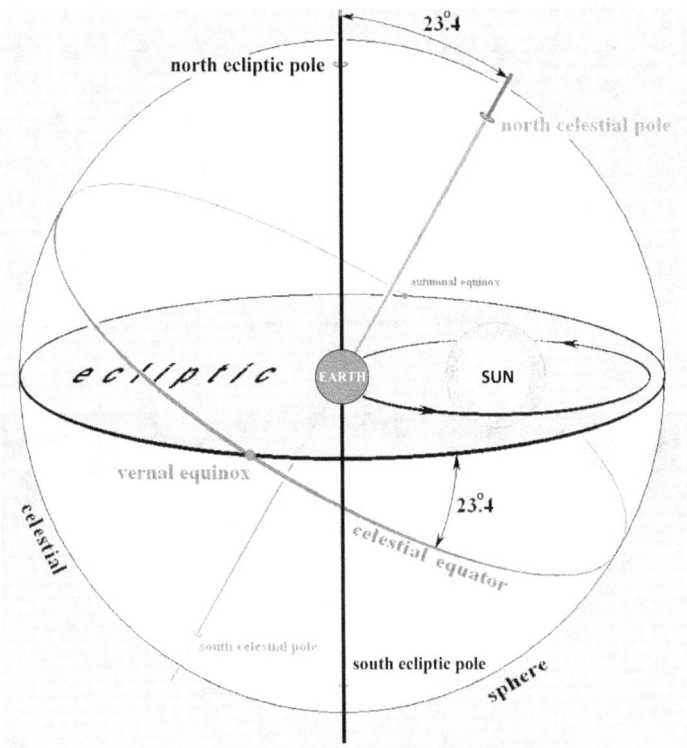

Figure 10.2 The plane of the ecliptic

"If that's the case," said Seb, "then the universe shouldn't look the same in all directions, right?"

"Exactly. We should see that the universe has a structure defined around the teli. And maybe we have!"

"What do you mean?"

"Well, the remnant radiation from the Big Bang is known as the cosmic microwave background, or CMB."

"You mentioned this when we were talking about fine-tuning, and the scientists who accidentally detected it because of pigeons."

115

"That's right. CMB represents a large-scale view of the universe, so it can be used to identify whether our position or movement has any significance—in other words, whether it's unique. It turns out that scientists studying the CMB have found unexpected results. Some large features of the microwave sky at distances of over thirteen billion light years appear to be aligned with both the motion and the orientation of the solar system.[12] For example, the top half of the CMB—by top I mean with respect to the plane of the ecliptic—is cooler than the bottom half. And there are other, more complex side-to-side features. It's still not clear whether this is due to systematic errors in processing, or contamination of data by local effects, or an unexplained violation of our assumption that we don't occupy a special place and the universe is the same in all directions. But these anomalies keep showing up in repeated satellite measurements with ever-improving technology."[13]

Figure 10.3 The anisotropy in the cosmic microwave background

One anomaly is an asymmetry in the average temperatures on opposite hemispheres of the sky (indicated by the curved line), with slightly higher average temperatures in the southern ecliptic hemisphere and slightly lower average temperatures in the northern ecliptic hemisphere. There is also a cold spot that extends over a patch of sky that is much larger than expected (circled).

Chapter 10 – It's About Time

"Wow, that's really interesting." Seb lay back, propped up by his beach bag, and clasped his hands behind his head. "So what are cosmologists saying?"

"They're undecided. But give me a sec while I Google how a respected cosmologist summarized it." It took me only a few moments to find the passage, then I read it out loud:

> But when you look at the CMB map, you also see that the structure that is observed, is in fact, in a weird way, correlated with the plane of the earth around the sun. ... That's crazy. We're looking out at the whole universe. There's no way there should be a correlation of structure with our motion of the earth around the sun—the plane of the earth around the sun—the ecliptic. That would say we are truly the center of the universe.[14]

"The correspondences and differences are fascinating," Seb said. "So, it seems there's agreement that the universe has expanded from a small space, but beyond that, religion and science are still at odds about where that small place came from, whether the expansion is still going on, and whether we occupy a special place in the universe."[15]

I nodded. "That's where things are at, yes."

"OK, one last question; I've heard that something called string theory proposes ten or eleven dimensions, not just three of space and one of time. What does the Torah have to say about this?"

"You're right. We're immediately aware of the three dimensions that surround us on a daily basis, the three spatial directions that define the length, width, and depth of all objects in our universe. Next, we have the time dimension. That's all, as far as we can perceive. The possibility of more dimensions arises because some proposed theories of physics aimed at combining quantum mechanics with gravity need ten or eleven dimensions. However, the Kabbalistic text describing the complete Standard Model, which we'll focus on next, clearly states in its first verse that there are five dimensions: universe, year, and soul.[16] In modern terms, these translate to the three spatial dimensions, the time dimension, and the soul dimension; the last refers to our ability, due to free will, to choose between right and wrong at every point in

117

time and space. This is a direct result of the fundamental design of Creation. We saw that God created the universe via His ten channels of divine energy, the sefirot. These ten correspond to the ten ends of each of the five dimensions. This is how the sources put it," I said, reading from my iPad:[17]

> The Sefirot of Nothingness: Their measure is ten, which have no end
> A depth of beginning and a depth of end [time]
> A depth of good and a depth of evil [soul]
> A depth of above and a depth of below [space]
> A depth of east and a depth of west [space]
> A depth of north and a depth of south [space]

Seb exhaled, then smiled. "Heavy stuff. But lunch has given me more energy again. Can we review time now, how the Torah's account of time compares with science's?"

"Sure. The explanation isn't as lengthy as for space. The Torah tells us that once we have space and particles, we can count something, and that this counting, this measured motion, is physical time—very mysterious but very simple and in complete agreement with scientific observation."

"Yeah, we talked about that yesterday," he said, "that science defines time by counting the cycles of something—like a pendulum, or electrons changing energy levels in an atomic clock—but it doesn't actually *measure time*."

"That's right," I said. "Interestingly, the Torah goes much further in explaining time. Like everything else, time goes from the general concept or category to the particulars or details, with nothing in the particular that doesn't also exist in the general."[18]

Seb scrunched his face. "What are you getting at?"

"What I mean is that time is the physical manifestation of a spiritual counterpart called the *order of time*. This order of time establishes the full history of the universe, with its various phases, according to the biblical timeline. It introduces an order, a flow from beginning to end. This flow in one direction, to the future, is what science calls the arrow of time. Biblically, the arrow of time is a design

Chapter 10 – It's About Time

of history from beginning to end. Also, although at a spiritual level the order of time is absolute, in the universe, physical time—the measurement of motion—is influenced by factors that influence the motion, such as speed and the strength of the gravitational field."

"That's what Einstein cleared up with his theories of relativity, right?" asked Seb.

"Right. In our world of classical physics, we see every clock keeping the same time, whether we're standing still, riding a car, or traveling on a plane. But Einstein showed that clocks that move keep time slower. The effect is small at our normal speeds, even flying on a plane, but as the speed of the traveling clock approaches the speed of light, it really slows down, and at the speed of light, the clock stops. In other words, his theory showed that time as we measure it slows down with speed and doesn't pass at all when traveling at the speed of light. This is actually found in the Torah!"

"Where?"

"The sources say that every physical creation has a spiritual precursor from which it derives its existence. In addition, its physical characteristics are synonymous with those of its spiritual source.[19] We don't know all the spiritual precursors, but we do know the spiritual precursor of light: the Ohr Ein Sof, the Godly light.[20] God is, by definition, above time. In other words, the physical light that derives from the Godly light doesn't experience time. This is the essence of Einstein's theory of special relativity."[21]

"Ah, OK!" His eyes lit up with understanding. "So, Torah and science agree on what physical time is and what affects it?"

"Yes, but the Torah also says that there is an order, a flow from past to future. Science is still grappling with explaining this directionality of time, the arrow of time."

"Phew." He looked alert, and I could tell his mind was processing. "I'm really glad we've tackled time and space. Do we take on the forces of nature and the elementary particles next?" I nodded. "I need to process first." He grinned. "How about we get dinner later and watch a movie back at the hotel?"

119

Chapter 11

Forces to Be Reckoned With

When dawn broke the next day, it revealed a dense fog. The forecast online said it might linger for much of the day, so rather than going out for a jog, I got a glass of milk, then settled in a comfortable chair and took out my iPad to read.

"What're you reading, Dan?" Seb asked sometime later, making me jump. He smiled. "Either it's something really absorbing or your hearing is going. You didn't even notice me come in."

"I'll opt for the first explanation," I replied wryly, stretching and getting up to make Seb a coffee. "It's definitely a mental workout, but fascinating. Aryeh Kaplan's translation of the *Sefer Yetzirah*, along with commentary."

"Aryeh Kaplan…" Seb squinted. "I remember you telling me about him a couple of years ago, on the kayaking trip. He was an American rabbi who translated a really rare text found in a Russian library." He looked sheepish. "I can't actually remember more than that right now."

"I'm impressed you remembered that much."

"Remind me about him," Seb said, flopping on the couch and starting to peel a banana.

"Well, up until age thirty-one, he was a successful physicist. He had been raised in a nonobservant Jewish home, so he didn't start learning Hebrew until his early teens. In fact, after his mother died, when he was thirteen, he ended up getting kicked out of school for 'acting out.' Fortunately, rather than losing his way as a street kid in the Bronx, he was befriended by a boy his age and began studying at a yeshiva. Although he continued his religious studies in Israel, he ended up returning to the United States to pursue bachelor's and master's degrees in physics. But after about four years as a leading research scientist, he switched careers and became a rabbi."

Chapter 11 – Forces to Be Reckoned With

Seb leaned forward. "I remember that he was an incredibly prolific writer, right?"

I nodded. "Nearly fifty books by age forty-nine, when he died of a heart attack. His translation of and commentary on the *Sefer Yetzirah* makes it accessible to English readers. The *Sefer Yetzirah* is a very important book in the history of Kabbalah, and it's key for what we were planning to talk about next."

"The *Sefer Yetzirah* contains information about the forces of nature and elementary particles?" Seb's eyebrows went up.

"Yup. That's why I was reviewing it this morning."

"OK, I'm ready when you are," he said, stretching his arms above his head, then kicking back on the couch. "Can you tell me more about that book?"

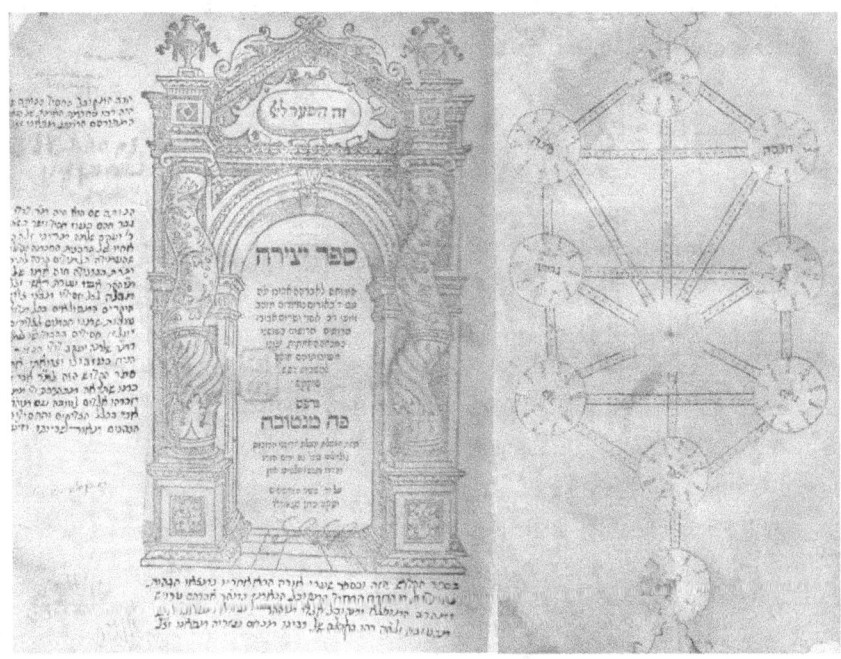

Figure 11.1 *Sefer Yetzirah*, 1562: cover and inside page depicting sefirot diagram

I nodded. "Let me read you some highlights from Kaplan's introduction." I skimmed the screen. "The *Sefer Yetzirah* is the oldest Kabbalistic text. References to it date back to the first century, text from it is quoted in the sixth century, and the first commentaries on it that we know of appeared in the tenth century. The title translates to the Book of Creation. Yetzirah is a Kabbalistic term referring to the third of the four spiritual worlds, the realm where created, limited beings take on their form and definition. So, it's where we can try to understand the state of the physical creation out of nothing."

"Who wrote it?"

"Because it's so ancient, historians can't determine its origins, but tradition attributes it to Abraham, the progenitor of the Jewish nation and common patriarch of Christianity, Islam, Judaism, and other religions."

"So how did we end up with the text we have today?"

"Originally, it was passed on orally, exactly word for word, with the leaders of academies keeping written notes to ensure the tradition was accurately preserved, but these notes were never published. After many centuries, these notes were expanded to include the whole text, and this was then redacted and published by Rabbi Akiva."

Seb snapped his fingers. "That's the famous scholar who supported the Bar Kokhba rebellion against the Romans in the second century, right? We talked about him back when you were working on *The Biblical Clock*."[1]

"Yes, one of the greatest scholars of all time."

"Passing it on word by word sounds incredibly difficult," he remarked. "I had to memorize all of Hamlet's soliloquies in high school one year, and it took me ages. But a whole book?"

I smiled. "Well, keep in mind that in cultures based on oral tradition, people start learning how to memorize the spoken word from childhood, so their brains are more geared toward that kind of memory feat. But also, the *Sefer Yetzirah* is only about 2,000 words long."

"That's the length of a short essay! How can it contain so much wisdom?"

"First, it's not like a normal secular book that introduces a topic, then provides all the background, then follows a set of logical arguments to elaborate on the topic and reach certain conclusions. This book is a divine record, so the information in it is presented with no substantiation or explanation. Second, it's intended to be understood only in the context of the full oral and written traditions, so it doesn't recapitulate or reference any of that; the reader is expected to have adequate prior knowledge and to do the necessary work to understand the contents."

"OK, that makes sense," he nodded.

"At the beginning of an ancient manuscript of the book, the scribe wrote: 'This is the book of the Letters of Abraham our father ... and when one gazes into it, there is no limit to his wisdom.' The gazing here means not just physically reading the book but meditating on it to gain insight. So the text is essentially meaningless to anyone who doesn't have an extensive background in the Torah. Fortunately, Kaplan did all the work for us novices and generated a translation, including most of the commentaries on the *Sefer Yetzirah*; his book is just under 400 pages!"

"Who else is the book relevant for?"

"Well, Abraham is the patriarch of many religions, so the book is important to them too, and has been for a long time. For example, although it was first printed in Hebrew in 1562 in Mantua, it had already been translated into Latin earlier by the Christian mystic Guillaume Postel and printed in Paris in 1552."

"OK," Seb said, wrinkling his forehead. "What does it talk about?"

"Its main topics are cosmology and cosmogony, and how the Divine accomplishes and sustains the Creation."

"Cosmogony—what's that?"

"The description of the creation or origin of the universe. The book is dominated by the role of the Hebrew letters as the building blocks of the world."

"Ah! So, we're getting to the building blocks." Seb rubbed his hands. "Wait. Can we quickly review the tohu and tikkun concepts first? I think that'll help me as we look at what the Torah says about forces and elementary particles."

"Sure. So, you'll recall that we have two fundamental existences: one left over from the world of tohu and one from the world of tikkun.[2] The classical physics of the macroscopic world belongs to the world of tikkun,[3] where everything is in vessels, constrained within time and space, and makes sense to us. The quantum microscopic world belongs to the world of tohu, where the microscopic particles aren't encased in vessels, so they're in perpetual motion, unconstrained by time and space, and things in that world don't make sense to us."

"OK, so the world of tohu corresponds to the microscopic described by quantum mechanics, and the world of tikkun to the macroscopic described by classical theories like general relativity?"

	World of tohu	World of tikkun
What it corresponds to	Microscopic world	Macroscopic world
How it came to be	Created	Made
Status	Perfect, unchanging, eternal	Changes, decays, is eventually reduced to its original components: tohu
What it consists of	Spark fused to vessel fragments	Sparks fused to vessel fragments concealed inside macroscopic objects (as if trapped in a crystal)
Limitations	Cannot be at rest; not constrained in time and space	Can be at rest; constrained by time and space
Life force	Life force is fused to it; permanent, never stops, perfect, lasts forever	Life force is inside the vessel; in living matter, it can leave at any time, and the container (body) dies and then decays
Lego analogy	Lego block, so small it is quantum	Many Lego blocks assembled into macroscopic objects

Figure 11.2 Characteristics of the worlds of tohu and tikkun

Chapter 11 – Forces to Be Reckoned With

"Right."

"And in the world of tohu, God created the building blocks of nature from nothing, whereas in the world of tikkun, everything was something made from something else, according to the laws of physics."

"Right again."

"And by definition, the things that were created—I mean created out of nothing, as a result of a bara event—are perfect, unchanging,[4] and everlasting.[5] And the forces and elementary particles we're going to talk about are perfect, unchanging, and eternal, right?"

"Not quite. They're perfect. The forces are eternal, and so are the lowest-level particles that everything is made of, but the higher-energy particles aren't eternal in the way we think of it, because they exist unchanged but they go in and out of existence and decay to lower-level particles. While they're all unchanging now, which ones have *always* been unchanging is an interesting question, one that physicists are asking too. For example, they wonder whether the speed of light has always been what we observe today. If it was different in the past, that could explain certain phenomena. Remember we were talking yesterday about the expansion of space coming to a stop at the end of Day 6 of Creation, 5,781 years ago? The sources actually say that everything in Creation was able to evolve and change—even what arose out of a bara event—from whenever it came to be until that point.[6] So it's possible that some of the forces' properties—strength, for example—and particles' properties, such as the speed of a photon of light, weren't constant and only became set in today's values 5,781 years ago."

"Got it," said Seb. "So moving on, everything that's made and exists in the world of tikkun changes, decays, and eventually returns to the original tohu components?" I nodded. "And because the tohu pieces consist of sparks attached to pieces of the broken vessels, they're not constrained by time and space, and we get those weird observations in quantum mechanics experiments. But in the world of tikkun, the sparks are concealed in larger macroscopic structures that are constrained by time and space, and we can make sense of them with classical mechanics."

"Yes, that's correct."

He exhaled for several seconds. "Could you remind me again about the concept that everything has a life force?"

"Sure. When the letters of the Hebrew alphabet appeared in their most material form, which was the elementary particles, these contained a life force or spark of the divine. This spark makes them everlasting. Coupled with the fact that they're not constrained by time or space, this means they don't stop moving; in fact, they're like perpetual motion machines."

Seb shook his head. "Wow. That explains why something like an electron spins forever at the same rate, which scientifically doesn't make sense. It should slow down over time, right?"

"Right, but it doesn't—something science hasn't explained!"

"And," added Seb, "the Torah descriptions of the worlds of tohu and tikkun match what scientists see with their instruments."

"Correct. Many questions relating to these two worlds remain unresolved in science. But the Torah sheds light on them, particularly on the most pressing problem of what happens when these two worlds meet. In science, this is known as the problem of interpreting quantum mechanics to explain how it corresponds to reality. Although quantum mechanics has withstood rigorous and thorough experimental testing, many of these experiments are open to different interpretations of what's actually going on.[7] A few interpretations of quantum mechanics have been developed, but none is totally satisfactory or fully accepted. Right now, the most accepted one is the Copenhagen interpretation, although many objections to it have been raised since the mid-1920s, when Niels Bohr and Werner Heisenberg proposed it. But the Torah's explanation of what happens when the worlds meet is simple and I think answers many of the objections to the various interpretations."

"Simple sounds appealing," Seb said, "but let's not get into that right now. We'd better move on to the forces and particles; we won't be on holiday forever!" and he grinned.

I glanced outside. The fog remained thick. "Sounds good." I opened a bottle of orange juice, poured us both a glass, then got comfortable.

Chapter 11 – Forces to Be Reckoned With

"The Torah elucidates the four forces of nature, which you'll recall are the strong nuclear force, the weak nuclear force, electromagnetism, and gravity. In science, we've simply discovered these forces. We don't know why there are four, or why they have their specific properties."

"Does the Torah tell us this?" he asked, looking at me intently.

"Well, I've done a lot more work on particles than on forces, but the little I've done certainly tells us a lot. For example, these forces emanate from God's essential name, which consists of four letters.[8] We can therefore be certain there are only four forces."

"But isn't science certain about that?"

"Some scientists have proposed the possibility of another force relating to unexplained phenomena.[9] The Torah is clear there are only four. It also tells us something about each force." I opened up a file on the iPad and went to sit by him on the couch.

Symbol	Name	Connotation	Force
י	Yod	Infinitesimal point Very short range	Strong
ה	Hei	Infinite Filling of all space	Electromagnetic
ו	Vav	Very short connection or hook	Weak
ה	Hei	Infinite Filling of all space	Gravity

Figure 11.3 The letters of God's name and the forces of nature[10]

"We can learn the basics by looking at the shape of the letter for each force. The first letter in God's essential name is yod.[11] It's a dot but drawn as a very small symbol that looks like a comma, and it connotes an infinitesimal point, corresponding to a very strong, short-range force."

"The strong nuclear force?"

"That's right. The second letter in God's essential name is hei."

"There are two of those in His name," Seb remarked.

"Yes, and we'll come to that in a moment. The hei[12] represents the filling of all space. It consists of a vertical line attached to a horizontal line and a little stub. Its vertical side and top horizontal side indicate those two directions in space as written on a page; the little stub in the middle denotes the direction into the page. In this way, the letter describes a force that fills all space and is infinite in range."

Seb glanced sideways at me. "That sounds like gravity."

"Yes, it does, and gravity is the final hei in God's name, but the first hei is the electromagnetic force, which, like gravity, is infinite and fills all space. The third letter, vav,[13] which looks a bit like a walking cane or a hook, indicates a very short connection."

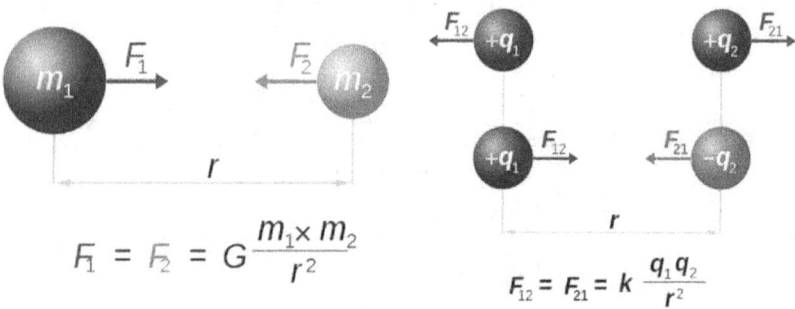

Figure 11.4 Newton's law of gravitation and Coulomb's law of electrostatics have almost identical forms and act over infinite distance

"Sounds like the weak nuclear force," he said, and I nodded. "That *is* amazingly simple."

"Yes. I'm sure there's more information in the Torah on the forces, but I've focused on particles to make sure this information is derived before CERN discovers more of them!" I glanced outside. "Hey, the fog is starting to lift. Should we get down to the beach for a few hours of surfing?"

"Great idea," he said, standing up from the couch and stretching his back. "It'll be good to get some exercise before we move on to the letters and particles."

Chapter 12

The Signature of God

The thick morning fog apparently had prompted many people to change their plans for the day, as when we got to the beach, only a few surfers were paddling out to meet some fabulous waves. We spent several exhilarating hours out there, stopping only for a short snack break on the beach to refuel. Since our flight back to Vancouver was leaving midday tomorrow, this was our last chance to surf, so we made the most of it.

Returning to the hotel, exhausted and happy, we decided to get cleaned up and then continue our earlier discussion back in my room.

I was reading the *Sefer Yetzirah* and downing a second glass of juice when Seb walked in, exclaiming, "I'm ready to get into letters now. Are you reading the Book of Creation again?"

"Yes. I was looking at what it says about the letter tav and comparing it to what many other sources say about it."

"Other sources? You mean there are books that talk about the Hebrew letters?"

"There certainly are. Like I was telling you earlier, the shape, sound, and numerical values of the letters provide deep meaning, so numerous sources have expounded on them. The Talmud states that tav (ת), the last letter in the Hebrew alphabet, represents the word emes, אמת, meaning truth. The reason emes is represented by its last letter and not its first, alef (א), is that the essence of truth is determined at the end of a journey or passage, not at the beginning. Often when we begin something, the truth of the matter doesn't seem attractive. Only once we see the outcome do we appreciate that the path of emes was the only way to travel."

Seb nodded, then raised his eyebrows, asking me to continue.

"What I'm interested in is where the Talmud states: 'The signature of God is the word emes.' Just as painters put their signatures on their paintings, God imbeds His signature in the universe, His creation. This concept is found in the Zohar in its explanation of the last three words in the story of Creation: 'bara Elokim laasos [He rested from all His work] which God created and made.' The final letters of these three words (ברא אלקים לעשות) spell the word emes. Why? Because God's painting is imbued with His signature of truth."

"So what's that got to do with physics?"

"Well, I've discovered through other sources that tav is the letter corresponding to the particle that carries the force of gravity. But we're getting ahead of ourselves here."

Seb smiled. "You're right. OK, let's start at the beginning of the particle story."

"We've seen that everything is made by combining the twenty-two entities referred to as the letters of the Hebrew alphabet. During the creation process, these go from the general concept or category to the particular, the details. At the lowest, most basic level of reality, where we exist, they correspond to the particles of nature, so we'd expect there to be twenty-two elementary particles.[1] Kabbalah also gives us the exact arrangement of these letters—in effect, its version of the Standard Model. This was presented in the *Sefer Yetzirah* almost 4,000 years ago."[2]

"Amazing."

"Right? Here, let me show you a diagram in the *Sefer Yetzirah*. I've annotated it to make the comparison to the Standard Model easier."

Chapter 12 – The Signature of God

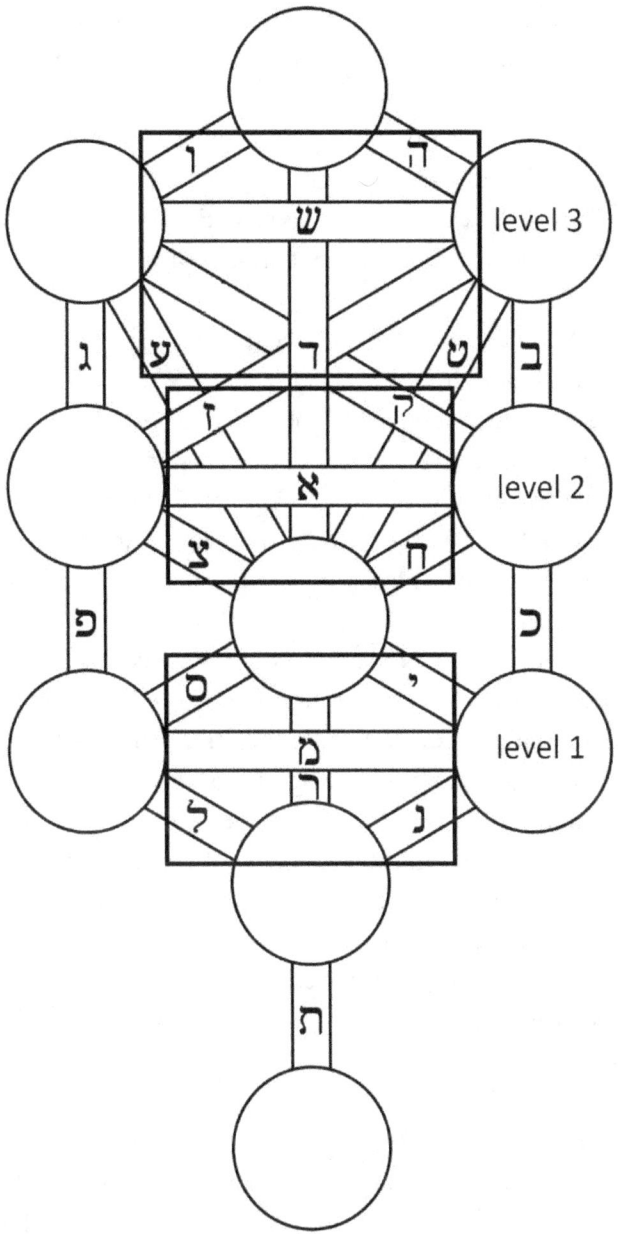

Figure 12.1 The Hebrew letters and the elementary particles as defined by the Arizal

Seb peered at the iPad screen. "That sort of looks like a person, with a torso in the center, and a left and right side that mirror each other."

"Yes. In fact, the human body also corresponds to this diagram.[3] The circles in the diagram are the ten sefirot, which are the ten emanations through which God creates and interacts with the world. These ten can be connected with twenty-two lines—no more, no less. That's just math."

"Wait," Seb said, pointing to the screen. "Are those connection lines the twenty-two letters?"

"Yes. This is the main part of the *Sefer Yetzirah*, the ten sefirot, the twenty-two letters, and how they underlie and produce all reality."

"Hey," he said, some excitement creeping into his voice, "I notice the diagram and therefore the letters are organized in three levels. Does that relate to the fact that the elementary particles come in three energy levels?"

"Correct," I smiled. "Now, remember: science has no explanation for why there are three levels. But here, the three levels are part of the design of everything. They flow from the arrangement of the sefirot as spelled out in the *Sefer Yetzirah*—4,000 years ago."

Seb shifted to look more closely at the iPad. "When I look at how the ten circles representing the sefirot are connected by the twenty-two letters, the first thing I notice is three horizontal connections, one at each level. What's that about?"

"These are the three mother letters. The *Sefer Yetzirah* calls them 'mother' because they're primary: one is the first in the alphabet, one is the middle letter, and one is the second to last letter."[4]

"What do they represent?

"The three particles that carry forces.[5] The top letter corresponds to the photon, which carries the electromagnetic force, and the bottom one corresponds to the strong force particle, which is the gluon. The middle letter corresponds to the weak force particles."

"Wait a minute. You told me before that the weak force is carried by three particles."

I smiled. "Good memory. The letter that carries the weak force, the alef, is actually a composite of three letters: a yod at the top, a vav tilted on the diagonal, and an inverted yod below that."

Chapter 12 – The Signature of God

"That matches the w and z bosons we saw in the CERN video!"

"Exactly. There are two w bosons, identical but opposite in charge. Here, we have two yod that are identical, but one is upside down. Then we have the vav, corresponding to the z boson."

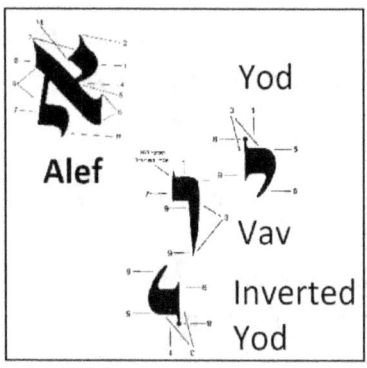

Figure 12.2 The letter alef, made of three letters

"Do the shapes of the other letters say anything about those forces?"

"Definitely. For example, the mem that carries the strong force actually comes in two forms: one that's used at the beginning and in the middle of words (מ), and one used at the end of words (ם). Each of the two forms is drawn from four separate but almost identical parts that are put together to make the letter, so in total, eight parts are used to draw the two mems. And how many types—or more correctly, color states—of gluons do you think there are?"

"Eight?"[6]

I nodded. "So, what do you see next?"

"I see twelve diagonal lines of connecting letters, with four at each of the three levels—two originating from the left and two from the right. What are they?"

"The *Sefer Yetzirah* calls them the twelve elementals, which result from the twelve possible permutations—meaning arrangements—of the four letters in God's essential name.[7] They represent the constituents of all visible matter."[8]

"How so?"

133

"We saw that the particles of matter come in three energy levels, and at each level there are four particles: two quarks and two leptons. We see the same here—three levels with four particles each. The two on the left are quarks, and the two on the right are leptons."

"Why?"

"Because in the conception of the sefirot, left always signifies bounded and constrained, indicating quarks are constrained. And this is accurate, as we don't see them alone; they always group to make something, such as a proton. And right always represents unconstrained, like the electron and muon, which can go anywhere on their own."

Seb crossed his arms and looked thoughtful. "OK, so we've explained the three forces and the twelve particles of matter. That's the Standard Model, right?" I nodded. "Except for one more particle, the Higgs boson?"

"Yes. Now, look here—what else do you see?" I pointed at the screen.

He peered for several moments. "I see seven vertical lines of connecting letters: two on the right, two on the left, and three in the middle. So, what's their significance?"

"These seven vertical connecting letters are what the *Sefer Yetzirah* calls the seven doubles—doubles because each of these letters carries two sounds, like ph and f.[9] One of these seven corresponds to the Higgs boson, and the other six are the missing six particles in the Standard Model."

"Wow, that's elaborate. Can you explain a bit more?"

"Sure. There are seven letters arranged vertically. Vertical means interacting with gravity, not with the three forces carried by the horizontal letters. Two letters are on the left, so they're constrained; these correspond to heavy dark matter particles. Two are on the right, so they're unconstrained and correspond to light dark matter particles. One of the vertical letters, the lowest one, is in the middle, and that's the final letter of the alphabet, the tav that we discussed earlier. The tav sustains the whole structure and is connected to the sefirot above. Its meaning is seal or stamp, and it represents the base of the world. It corresponds to the graviton, the particle for the force of gravity that holds the universe together. The next one above it is the Higgs boson, and the highest is a Higgs-like heavier partner. Note that this letter is almost identical in shape to the letter for the Higgs."

Seb held up one hand. "So far, we've talked about the form of the letters and their position in the diagram, but each Hebrew letter also has a numerical value. Do those numbers play a role too?"

"Yup. I could show you that the numerical values of the letters, coupled with their position on the diagram, allow us to calculate the mass of many of the particles."

"You mean we can predict the mass of the undiscovered particles?"

I nodded. "For example, the mass of the lowest, and therefore most stable, heavy dark matter particle is about sixty percent the mass of a proton, and the mass of the lowest light dark matter particle is about twenty times the mass of an electron. But that's an involved story that we're not ready for now."

"OK, then getting back to the three-level organization," said Seb, "what else can we learn from it?"

"Kabbalah tells us that only the letters—that is, the particles—in the lowest level of the diagram have infinite life and make up everything. So, in the diagram we've discussed and in the Standard Model, that's the two lowest-energy quarks, the electron, and the neutrino.[10] The particles in the higher levels of creation come and go out of existence and decay to other particles, meaning they have finite lifetimes."

"That's what science has discovered too," he said. "But I thought there were other particles with infinite life, like the photon, and yet that force-carrier particle is on level three, not level one."

"You're absolutely right, yes; there are other particles that we think have infinite life, but all those are massless."

"Right, the photon has no mass and travels at the speed of light."

I nodded. "And what happens at that speed?"

"Time doesn't elapse."

"Exactly. So, the photon might have a lifetime of a fraction of a second, but it's traveling at the speed of light, so it experiences no time and, in our eyes, lasts forever."

135

He rumpled his hair. "Then the particles at the lowest level have infinite life, and those at other levels that have no mass, like the photon, *appear* to us to have infinite life."

"Correct."

"But what about the dark matter particles on the right and left that seem to be between levels?"

"Well, I think the lowest-level dark matter particles, which are between the first and second levels, may not be permanent, because they're not purely in level one; so, they have a finite life. In fact, some scientists have conjectured that dark matter may decay."[11]

He sat back heavily against the cushions and exhaled. "It's amazing that the form and function of the whole Standard Model, even the undiscovered particles, was revealed four millennia ago. It'll be exciting to see if all these predictions match when the dark matter particles are discovered!"

"For sure. There's no doubt about the number of particles and their features, but I'm still refining the mass predictions for the undiscovered particles."

"Hang on," he said. "You've left out something major."

"What?"

"Antiparticles, antimatter. Where do they come in?"

I laughed. "Good point! They're one of the big unsolved issues in physics." He motioned for me to continue. "Antiparticles have the same mass as their corresponding particles, but qualities such as electric charge are opposite. Matter and antimatter are therefore always produced as a pair. So according to the Big Bang theory, at the very beginning, during the first fraction of a second, the hot, dense universe was buzzing with equal amounts of matter and antimatter. Herein lies the problem. If there were equal amounts of each, they should all have combined again, cancelling each other out, and we shouldn't exist. Yet calculations show that one in a billion particles of matter managed to survive, and from them the universe developed. This is known as the matter–antimatter asymmetry problem.[12] Neither the Standard Model of particle physics nor general relativity provides an explanation for this. There are some hypotheses, but none are widely accepted."

"What does the Torah say about this?"

"I'm still working on it, but one amazing correspondence is clear. The Torah agrees that the early state was a chaos of particles—the world of tohu. It also says that the particles devolved from the twenty-two Hebrew letters."

"So…?" he asked.

"Well, the letters can be written right to left or left to right. A letter written left to right is the 'anti-letter' of the same one written right to left."

"Makes sense."

"So, when God chose to continue with Creation by creating our world—the world of tikkun—He picked only right-to-left letters from the chaos. You see, he formed 'words,' combinations of letters, which make up everything that exists, only from right to left, as the Hebrew language is written, meaning He picked particles, not antiparticles, to make everything."

"But where is the amazing correspondence?"

"The first letter in the Torah is beit, the second letter of the alphabet. It looks like a square c back to front (ב), with all sides closed except the left side, where any subsequent text continues. This is explained as showing that the past is closed and the history or Creation narrative proceeds only to the future.[13] Thus, writing right to left is equivalent to time flowing in one direction only. The Torah is therefore telling us that the matter–antimatter asymmetry and the direction-of-time problem are connected. We have a world made of matter because time flows in one direction."

"Interesting… Has any scientist conjectured this?"

"Yes, some have investigated it.[14] And it's been assumed for a long time that an antiparticle is indistinguishable from the mirror image of its particle running backward in time."[15]

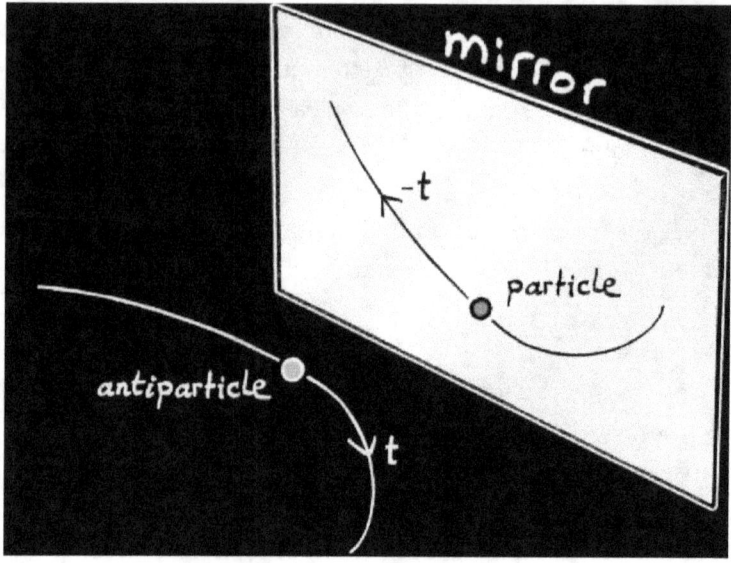

Figure 12.3 Particle and antiparticle

An antiparticle is the mirror image of its particle running backward in time.

At that point, Seb's stomach rumbled loudly, and we both laughed. "OK," he said, "I'm out of questions, for now at least."

"Let's get something to eat," I suggested. "Then we should pack so we're ready to leave early in the morning. San Diego Airport is an hour away."

As we got up to head out for dinner, Seb said, "Are you going to put all of this in a book? I mean, since I can understand your explanations, I bet a whole lot of other people could too. I think you should."

I smiled. "I just might do that."

Chapter 13

Summing Up

So far, we've explored knowledge about the development of the universe based on current scientific investigation and on Torah revelation. In particular, we have studied two periods of time: cosmic history and the first instant. This chapter summarizes the questions and findings so far.

Cosmic History

When it comes to cosmic history, we have seen that the biblical account states the universe will appear to have developed naturally—thus, the scientific method will be successful at explaining its development. Furthermore, the biblical account says things in the timeline for the development of the universe will appear to be the age that science has measured. There are only two exceptions to this agreement:

- The estimated age of the Earth based on scientific inquiry is about 4.5 billion years. However, **the biblical timeline says the Earth should appear to be about 7.5 to 8 billion years old.**

- Science and the Torah agree the universe has been expanding since the beginning, first decelerating and then accelerating. However, **biblical sources say the expansion of the universe stopped 5,781 years ago.**

The First Instant

When we look at the first instant, the Torah account clearly states that we will not understand part of it because it was a creation from nothing, which the scientific process does not deal with. As we study what that out of nothing creation entailed—how it happened and what it created—we find many answers to scientific mysteries, as well as many predictions about what we will discover. Here, we'll walk through the unknown or not yet understood issues we have unveiled in the prior chapters, and we'll summarize what the Torah says about them and predicts we will discover. Please note that this is not a complete list of scientific unknowns or mysteries[1] but only what we have discerned at the level of detail covered in previous chapters. However, the list does include many of the big questions.

Let's move logically through the questions raised in earlier chapters and recap both bodies of knowledge. The contents of this section are summarized in Figure 13.1.

- What came before the Big Bang?

    We saw that the Big Bang theory says nothing about time zero. We further saw that even if inflation is correct and we can get closer to time zero, we still need at least space and gravity to start inflation and the multiverse. The Torah predicts that the scientific method cannot reach an understanding of time zero or what came before it. The original start of the universe was an out of nothing creation that brought into being space, time, the elementary particles, and the forces of nature. From that point on, everything can be understood via the scientific method. What preceded this first instant was a spiritual plane, not a physical one, in which there is no physical time, only an order of time or sequential plan.

- Why does time flow in one direction only?

    We saw that there is no complete scientific explanation for the arrow of time. In the Torah, time was first created in the spiritual level as an order of time or a sequence of events—in one direction, from beginning to end. The whole history was foreseen in "one glance" and then was split up and subdivided. Once physical space was created and things in that space could move and change, physical time became

Chapter 13 – Summing Up

simply the movement or change of things, a counting of events. However, the flow of the development of the universe is governed by the order of time in one direction. When we measure how much time has elapsed by counting some movement or change, we simply record that flow (including any effects due to things that affect the counting, such as gravity).

- Why is time affected by motion and gravity?

Einstein's theories show that what we call time is affected by motion and by gravity. However, in science we don't have a definition of time, just a standard. The counting of a certain number of atomic-level vibrations is, by convention, what we call a second. In the Torah, physical time is only the measured motion of things in space, the counting of regular events such as the revolutions of the Earth, the swings of a pendulum, and the disintegrations of radioactive compounds. So, anything that affects the *rate* at which these events happen affects our counting and therefore what we call time. In particular, speed (rate of motion) and gravity affect these events that we count.

- Are time and space quantized or continuous?

We've discovered that most fundamental entities of the universe are quantized in discrete packets. Even light comes in photons of particular energy. Does this mean time and space are also quantized into tiny bits, or are they continuous—ever divisible into smaller and smaller bits—as we perceive them to be? In science, we simply don't know yet. The Torah says they are continuous. In the case of time, this is very clear: it was created in one glance and then subdivided into endless smaller pieces, but these are all part of the continuum. In the case of space, this is not as clear, but since the first space was millimeters in size and then stretched (i.e., expanded), it's unlikely to have been quantized. If it had been, it would have been created as one quantum and replicated, like a photon or electron.

- Did inflation happen?

  As we explored in Chapter 5, inflation has been proposed to solve a variety of problems. It also provides a method of "naturally" kick-starting the universe. However, we also saw that recent observations, if upheld, indicate the observable universe was never smaller than a few millimeters. The mystical tradition about the beginning of space says that a macroscopic piece of space was created; it doesn't require or talk about inflation. It also tells us that the expansion proceeded as a ball of thread would unravel. This description doesn't in any way allude to two very different periods of expansion (inflation and post-inflation), but it indicates continuous expansion, albeit first decelerating, then accelerating.

- What is the nature of the accelerating expansion of the universe?

  We have discovered that post-inflation, the universe expanded at a decelerating rate. For the past several billion years, though, it has been expanding at an accelerating rate. The theorized dark energy that drives this behavior is not understood; indeed, it hasn't even been discovered. The Torah says (1) the universe was created expanding and (2) the expansion follows the physics of an unrolling ball of thread that first decelerates, then accelerates. Unfortunately, I've not yet found a description of the mechanism for expansion, other than that it is part of the original out of nothing creation; in other words, when He "created the world, it went on expanding."

- Why/how are the twenty-five or more parameters that drive the Big Bang theory precisely fine-tuned to allow the universe to exist? And why can we only measure their values rather than compute them?

  The Big Bang is a very successful theory, but to drive it we require a large number of parameters that need to have precise, unexplainable values. Scientifically, it's hard to explain this other than by positing that (1) we need more time to make more discoveries that will yield an answer or (2) billions of universes exist, all of which started with a random set of parameters, and we live in the one universe that has the right set of values for us to exist. It's unclear whether this multiverse explanation can ever be scientifically tested. The Bible says that the beginning was an out of nothing, supernatural event, with the Creator setting the parameters so that the whole thing would work—not a satisfying answer for someone seeking a natural explanation, but an answer that, like a proper scientific theory, can be tested with the scientific method. How? By looking at what the Torah says about these parameters and seeing whether that matches what science measures. For example, the mystical tradition's prediction of the number and types of elementary particles that were created can be verified or falsified by scientific inquiry. Furthermore, the values of these parameters (for example, the masses of the particles) have to be in the mystical tradition somewhere, and more research can be done to extract them. I have gone some of the way, with very encouraging results, which I'll outline in Part 2.

- Why is the distant universe so homogeneous?

As we've learned, this is known as the horizon problem. Science can explain it with the theory of inflation, but this theory remains speculative. No other accepted mechanism[2] has been found that would have established homogeneity quickly enough for the universe to expand without having regions that were disparate from each other. The Torah explanation of the beginning does not suffer from this problem because it starts from a macroscopic amount of space that was homogeneous and very symmetric: God created space as a "very small point [and] left vacant an evenly measured place on all sides and expanding."

- Is there an anomaly in the cosmic microwave background that shows the Earth/solar system is at a special location?

Scientific theories assume that there is no special place in the universe, and thus the CMB should look the same in all directions. However, some large features of the microwave sky at distances of over thirteen billion light years appear to be aligned with both the motion and the orientation of the solar system (the plane of the ecliptic). It's still not clear whether these anomalies are real. The Torah is clear: the universe is centered and structured around the teli. The best description of the teli (predating any scientific findings) is that it is the plane of the ecliptic. So, we should expect to see structure organized around the teli.

- How does quantum mechanics give rise to reality? What is the role of measurement in quantum mechanics?

With quantum mechanics, we can make incredibly accurate predictions about the probabilities of events—for example, finding a particle at some point after it has gone through a particular opening. But we can only actually see or confirm these predictions by measuring the position of such a particle over and over again until we can calculate a probability of it being at any point and thus verify the theory's prediction. Think about tossing a coin. In our familiar world, we can predict that we will obtain heads half the time when we toss a coin. But we can only verify this by repeatedly tossing the coin. In quantum mechanics, we don't understand what happens when we measure the position of a particle. How does the probability turn into an actual position? Nor do we really know the status of the particle prior to the measurement. Was it in many places at once, and when we measured, we fixed it according to the predicted probability? The Torah says that quantum mechanics and classical mechanics are by nature separate realms. Microscopic particles described by quantum mechanics are unconstrained by time and space. The macroscopic objects in our everyday

Chapter 13 – Summing Up

life *are* constrained by time and space. Although microscopic objects can exhibit spooky behavior because they're unconstrained, the moment they interact with anything macroscopic—such as a measurement device—they become constrained and have known, definite outcomes. So the true "interpretation" of quantum mechanics is close to the Copenhagen interpretation, but no observer (conscious or not) making a measurement is required to end microscopic particles' quantum behavior; simply by interacting with a macroscopic object, they automatically become constrained by time and space.

- What is the meaning of non-local phenomena or entanglement?

We do not understand the spooky behavior of quantum objects—for example, the ability of particles to be entangled and affect each other at large distances simultaneously. Our classical world, with its speed limit of the speed of light, says this is impossible. Yet we continue to verify these effects experimentally. The Torah says that these microscopic particles are unconstrained by the notion of time and space and thus do not obey speed limits, which are by definition related to time and space. Once the particles interact with the macroscopic world, they become bound by time and space, and all these spooky effects stop.

- Why do some elementary particles apparently last forever, and why do all have permanent properties, such as mass, spin, and charge?

As best as we can tell, the particles from which matter is made seem to have infinite lifetimes. The electron's lifetime has been estimated to be at least five quintillion times the age of the universe.[3] Furthermore, if we left that electron undisturbed, it would spin at its rate forever and hold its charge. But this would be like a "perpetual motion" device, which physics says is impossible. Only these few building blocks of the universe last forever; everything else decays with time. Other elementary particles also decay, but their properties, such as spin, never change. These seeming contradictions are puzzling.

The Torah is clear that in the beginning, certain constituents are made out of nothing—space, time, the forces of nature, and elementary particles. The particles at the lowest level can manifest eternally; they can't decay or be destroyed. But everything else made from them can and will decay back to them. All particles with zero mass also exist forever because they don't experience time and can't decay. Although the higher-level particles are not eternal, they are still out of nothing creations, are perfect, and have constant, changeless properties.

- Why is light timeless?

Einstein's theory explains that as objects gain speed, the time they experience slows down. This is not just a measurement issue, it's reality. Elementary particles traveling near the speed of light live longer than the same particles going slower. When the speed of light is reached, time stops. So a photon, a particle of light, does not experience any time. It may have been traveling thirteen billion years to reach our telescopes, but it thinks it just left.

The Torah says there is a nonphysical existence, and everything physical is a manifestation of something spiritual after many devolutions. The Godly light—where all of time can be seen in one glance, where past, present, and future meet—is above time. Physical light is the physical creation that descends or devolves from Godly light and thus, like Godly light, does not experience time.

- Why does the Standard Model contain three generations of matter particles?

We saw that the Standard Model contains twelve matter particles: six quarks and six leptons. These come in three

Chapter 13 – Summing Up

generations or energy levels. In other words, the particles repeat as very similar particles, just with higher mass, three times (you may wish to refer back to Figure 6.2). Only the particles at the lowest energy level are stable; the rest have short lifetimes (although this is technically not true for neutrinos). There is no explanation of why we have three energy levels and not just one (or four, or two, etc.).

In the Torah, the complete Standard Model is a direct manifestation of the design of the universe around the ten sefirot, which are arranged in three levels (recall Figure 12.1). The matter particles in this model occupy three energy levels, and the model specifies that only the lowest-level particles (such as electrons) can manifest permanently in our physical universe. Dark matter particles (still undiscovered) exist in two levels that are in between the three levels of the matter particles.

- Why does the Standard Model have seventeen particles? What about the dark matter particles?

As illustrated in Figure 6.2, the Standard Model contains seventeen particles. We don't know why there are that many or why there are three generations of particles. We do know that this model is incomplete because it does not deal with dark matter, which doesn't interact with the electromagnetic force. This means it doesn't absorb, reflect, or emit light, making it extremely hard to spot.

The mystical tradition dating back 4,000 years contains a diagram of the complete Standard Model, comprising twenty-two particles representing the twenty two Hebrew letters with which the world was created. This diagram shows that the twelve matter particles come in three levels and two types (see Figure 12.1), as science has observed, and explains that only the lowest level is manifested permanently in the universe. The mystical model contains three force particles (for the forces, excluding gravity), matching the Standard Model except that the latter has two particles for the weak force, whereas in the mystical diagram, these are counted as one letter composed of the two particles. The mystical model includes two heavy dark matter particles and two light dark matter particles in two levels. These two energy levels exist between the three energy levels of the matter particles. The mystical model also contains a particle in the position to be the gravity force carrier (science's undiscovered graviton?) as well as a heavier, Higgs-like particle.

- Why do the particles have their specific property values?

The three generations of particles have specific parameters, such as mass, spin, and charge, that not only are unexplained but are fine-tuned for us to exist. The same goes for the force particles. We've discovered no complete relation between all these parameters. All we can do is experimentally measure every parameter for every particle.

The mystical tradition teaches that the properties of the particles relate to the properties of the corresponding letters: shape, sound, numerical value, and position in the Kabbalistic energy diagram (Figure 12.1). I have made progress in relating all of the particles' masses via the letters' numerical values and positions in the diagram, and we'll look at this in Part 2. But more work remains to fully uncover the information in the Kabbalistic diagram.

- What constitutes the universe's dark matter?

As discussed above, we don't know what dark matter is or how many particles of it there are. The complete Standard Model from the mystical tradition specifies four particles in two energy levels: two heavy and two light particles. The lower-energy heavy dark matter particle has a mass of about sixty percent the mass of a proton, and the lighter one has a mass of about twenty times the mass of an electron. The dark particles occur at energy levels between the three levels of the matter particles. Thus, the lower-energy dark matter particles

Chapter 13 – Summing Up

occur between the first and second energy levels of the matter particles (see Figure 12.1).

- What is dark energy?

Scientifically, we don't know what dark energy is. Biblically, I haven't found a direct reference to it. However, the expansion mechanism for the universe is part of the original out of nothing creation, as the sources say the universe was already expanding when it was created; this implies there is something driving the expansion that was created at the beginning. We are also told that the mathematics of the expansion relates to the mathematics of how balls of string unravel, which implies deceleration followed by acceleration.

- How many forces of nature are there?

We have discovered four fundamental forces of nature. Some scientists have proposed the existence of a fifth force. Physics holds that all four fundamental forces are related. The weak and electromagnetic forces have already been unified into the electroweak force, and good progress is being made in unifying this with the strong force. However, unifying these with gravity has so far proven intractable.

Biblically, the four forces of nature arise from God's essential four-letter name (the Tetragrammaton). Each force is represented by a letter, and each letter's shape relates to the fundamental reach of the force. Furthermore, these four forces unify together as part of the Tetragrammaton.

- How many dimensions does the universe have?

Dimensions are the different facets of what we perceive to be reality. We are immediately aware of the three dimensions that surround us on a daily basis, the three spatial directions that define the length, width, and depth of all objects in our universe. Next, we have the time dimension. That's all, as far as we can perceive. The possibility of more dimensions arises because some proposed theories of physics aimed at combining quantum mechanics with gravity need ten or eleven dimensions.

The Kabbalistic text describing the complete Standard Model clearly states, in its first verse, that there are five dimensions: universe, year, and soul.[4] In modern terms, these translate to the three spatial dimensions, the time dimension, and the soul dimension; the last refers to our ability, due to free will, to choose between right and wrong at every point in time and space.

The above information is summarized in Figure 13.1, for reference while reading Part 2.

Chapter 13 – Summing Up

Figure 13.1 Biblical elucidation of the first instant

Scientific question	Torah elucidation/prediction
What came before the Big Bang?	The scientific method will not be able to explain the first instant. What came before the Big Bang was God and then the creation out of nothing of space, time, particles, and forces.
Why does time flow in one direction only?	The order of time precedes the creation of anything physical. In it, all the universe's history is ordered from beginning to end. Once space is created, this order manifests in physical time proceeding in one direction to the future.
Why is time affected by motion and gravity?	Physical time is simply the measured motion of things in space. Thus, anything that affects the measurement of change or motion affects time—in particular, motion at high speed and gravity.
Are time and space quantized or continuous?	Space is continuous. It was created as one entity a few millimeters in size and expanding. Time was also created at one glance from the beginning to the end and then subdivided into smaller and smaller endless pieces. But it is continuous.
Did inflation happen?	No. Space was created at the macroscopic, millimeter size after what we call inflation, and there was never a singularity.
What is the nature of the expansion rate of the universe?	The expansion of the universe has always been occurring; it was part of the original out of nothing creation. It follows the physics of an unrolling ball of thread. Expansion first decelerates, then accelerates, as has been discovered.

Scientific question	Torah elucidation/prediction
Why/how are the 25+ parameters that drive the Big Bang theory precisely fine-tuned to allow the universe to exist? And why can we not compute but only measure their values?	The fundamental parameters of the universe were created out of nothing at the beginning, at their precise values. This process is by definition fine-tuned and not explainable by the scientific method. Some of the values of these parameters can be obtained from scripture (see Part 2, Chapter 17, Elementary Particles). Within the realm of science, we can only measure their values.
Why is the distant universe so homogeneous (creating the horizon problem)?	Space was created a macroscopic small point, leaving vacant an evenly measured, expanding place on all sides—a very homogeneous process. However, this process had a center and an alignment that led to some structure. (See the next response.)
Is the anomaly in the cosmic microwave background showing the Earth/solar system at a special location real?	The Torah is clear the universe is centered on the teli. It thus has a structure around this teli. The teli relates to the motion of the Earth around the sun.
How does quantum mechanics give rise to reality? What is the role of measurement?	Microscopic particles are unconstrained by time and space, so they can be in multiple places at once and influence the "past." The microscopic realm becomes constrained by time and space once it interacts with the macroscopic realm—whether this happens naturally or is forced by us, as when we make a measurement.

Chapter 13 – Summing Up

Scientific question	Torah elucidation/prediction
What is the meaning of non-local phenomena or entanglement?	Microscopic particles are unconstrained by time and space, so they exhibit the spooky behaviors we observe. However, when they interact with macroscopic objects, they are immediately trapped and become constrained by time and space, which ends their strange behavior.
Why do some elementary particles last forever, and why do all have permanent properties, such as mass, spin, and charge?	In the beginning, certain constituents were made out of nothing: space, time, forces of nature, and elementary particles. The particles at the lowest level are eternal and have unchanging properties, such as spin. They don't decay and cannot be destroyed—in other words, they can't return to nothing naturally. Everything else made from them can and will decay back to them. All particles with zero mass also exist forever because they do not experience time and can't decay. All twenty-two particles are perfect and have constant, unchanging properties because they are part of the original out of nothing creation.
Why is light timeless?	Light is the physical manifestation of Godly light. Thus, it moves at a speed at which past, present, and future are the same—time does not elapse. This is also where the quantum and classical worlds meet; objects at both scales can travel at that speed and not experience time's passage.

Scientific question	Torah elucidation/prediction
Why does the Standard Model contain three generations of matter particles?	The Standard Model is what we see of the essential design of all nature's building blocks. This design comes in three energy levels. Matter particles exist in all these levels, although only those in the lowest level can manifest eternally. The dark matter particles exist in between levels 1 and 2 and levels 2 and 3.
Why does the Standard Model have seventeen particles? What about the dark matter particles?	The number of particles in the world relate to the Hebrew letters with which the world was made. There are twenty-two elementary particles in the world. Each force carrier is counted as one, unlike in the Standard Model, where the weak force-carrier particles are counted as two. The known matter particles come in three distinct energy levels. Of the undiscovered particles, there is one more Higgs-like particle, the graviton, two "cold" dark matter particles (which are heavy), and two light or warm dark matter particles. These last four particles come in two distinct energy levels, not three like the rest of the particles.
Why do the particles have the specific property values they do?	Particle energies/masses are all related to each other as per the Kabbalistic energy diagram (Figure 12.1) and the Hebrew letters' numerical values, sounds, and shapes.
What constitutes most of the universe's dark matter?	Four dark matter particles: two heavy, constrained ones, and two light, unconstrained ones, each in two energy levels.

Chapter 13 – Summing Up

Scientific question	Torah elucidation/prediction
What is dark energy?	Space was created out of nothing, already expanding, and the expansion mechanism (dark energy?) is an out of nothing creation. We are also given an analogy that may describe the mathematics of expansion.
How many forces of nature are there?	There are only four forces of nature, per the four-letter name of God. Two have infinite reach, one is strong and extremely short range, and one has a connection that can easily be severed.
How many dimensions does the universe have?	There are five dimensions: three space, one time, and one spiritual. The spiritual dimension relates to our free will, our ability to choose between right and wrong at every point in time and space.

Part 2

In this part, we revisit some of the major topics and questions developed in Part 1 to explore them in more detail.

Each chapter in Part 2 is meant to stand on its own so the reader can read those of most interest. However, they are presented in a logical sequence if you wish to read them that way.

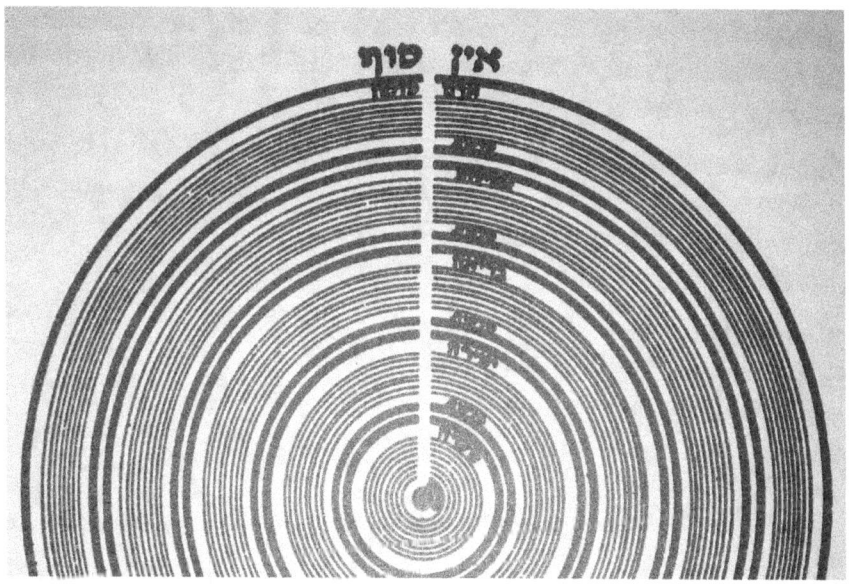

Stages of emanation and creation

Chapter 14

Before the Big Bang

We return to the question of what came before the Big Bang and how space, time, elementary particles, and the laws of physics emerged from that.

The Science Answer

We've seen that the scientific approach of working backward from the present day leads to a time near the beginning when the observable universe was very small and dense—after inflation, it was only a few millimeters to centimeters in size. At this point, everything required for the universe existed: space, time, elementary particles, and the laws of physics. As we go further back, with the aid of the theory of inflation, we can reach a point where the observable universe was even smaller, yet still a microscopic point of space existed, time existed, and at the very least gravity existed.[1] But we can go back no further.

Ideally, we'd like to go back to "nothing"—no universe and no laws of physics. However, physicists have no definition for this real "nothing"; in the context of physics, we cannot make sense of this sort of nothingness. We'd have to assume that there is such a thing as a state outside of time and space, where you can have the emergence of space–time from this hypothesized state of true nothingness. But how is that possible in the world of physics, which presupposes something? How does space–time emerge at a particular location when there's no such thing as space? How can you create the beginning of time if, without time already existing, there's no concept of "before"? And in this case, where would the rules governing particles and their interactions arise from? Does this real definition of "nothing" even mean anything at all, or is it just a logical construct with no physical manifestation?

Until the 1930s, the predominant scientific theory held that the universe was eternal. If the theory of inflation does not stick, and the observable universe we now live in must come from some small (but not microscopic) start, then science may return to an eternal view of the universe—perhaps the cyclical view. In this approach, the universe expands, then contracts, then expands, in endless cycles. The attraction of the cyclical model is that although the previous universe is squeezed down to a very small size, it is never at the scale where the most poorly understood quantum effects come into play. Further, it may be possible to show that the uniformity required at the start of our universe (which inflation tries to solve) arises naturally from the squeeze before the new or current universe expands again.[2] However, postulating an eternal universe just eliminates the question of how it all began, without answering it.

The physicist Sean M. Carroll has asked:[3]

> Do advances in modern physics and cosmology help us address these underlying questions, of why there is something called the universe at all, and why there are things called "the laws of physics," and why those laws seem to take the form of quantum mechanics, and why some particular wave function and Hamiltonian? In a word: no. I don't see how they could.

In short, science has no accepted answer to the questions of what came before the Big Bang and how space, time, elementary particles, and the laws of physics emerged from that.

The Biblical Answer

As we saw earlier, the biblical answer starts from the beginning and proceeds forward, relying on something eternal and nonphysical: God. It is God who at some point creates a start for the physical universe.

There is a story of a congregant who approached his rabbi to explain that he did not believe in God. Then he proceeded to describe God. The rabbi replied, "The God you don't believe in I also don't

Chapter 14 – Before the Big Bang

believe in!" So what God, what eternal being that got the universe started, does he believe in?

The Eternal—God

There is no way to define a complex being. Even when it comes to ourselves, the best we can do is talk about our behavior. More controversially, we can try to define our personalities. This difficulty is even more evident with God. So, let's summarize the biblical description of God, at least the most relevant parts of it for studying the beginning of the universe.

God Exists

The fact of God's existence is accepted almost without question. Proof is rarely offered. Genesis begins by stating, "In the beginning, God created…" It does not say who God is or how He was created.

God is a necessary prerequisite for the existence of the universe, and the existence of the universe is sufficient proof of God's existence.

God Is the Cause of Everything

Everything that exists is a product of God.

God Is Omnipresent

God is in all places at all times. He fills the universe and exceeds its scope. He is the God of all nations.

God Is Omnipotent

God can do anything. The only thing beyond His power is the fear of Him; that is, we have free will, and He cannot compel us to do His will.

God Is Omniscient

God knows all things, past, present, and future.

God Is Eternal

God transcends time. He has no beginning and no end. He is unchanged by anything that happens. He will always be there. When Moses asked for God's name, He replied, "Ehyeh asher ehyeh."[4] That phrase is generally translated as "I am that I am," but the word "ehyeh" can be present or future tense, yielding the additional translations "I am what I will be" and "I will be what I will be." The ambiguity of the phrase is often interpreted as a reference to God's eternal nature.[5]

God Is Incorporeal

Although many places in scripture speak of various parts of God's body (the hand of God, God's strong arm, etc.) or talk of God in anthropomorphic terms (God walking in the garden of Eden, etc.), it is a fundamental belief that God has no body or any kind of form. Any reference to God's body is a means of making God's actions more comprehensible to beings living in a material world.

This eternal, omnipotent being started the physical universe. As we have seen, He did so primarily by putting in place the system of nature, which we can study and use to get most of the answers about how the universe developed, but not how the very beginning happened. For that answer, we turn back to His design manual and review it in more detail.

In the Beginning

Most of the development of the universe, as narrated in Genesis and explained by scientists, proceeds from cause to effect to cause and so forth. For example, gravity causes denser areas of hydrogen to compress and become even denser. This in turn causes nuclear fusion to start, and a star is born, which causes… You get the idea. In science, we have been stuck, unable to determine the first cause, the cause of the beginning. In the Bible, we are told it is God. However, this initial cause is unlike all subsequent causes, because the physical universe is an out of nothing creation.

Chapter 14 – Before the Big Bang

Creating something out of nothing is a capability only the Creator possesses.[6] When the Midrash (a collection of non-legalistic teachings of the rabbis in the Talmudic era) discusses another out of nothing creation,[k] animals, it states: "if all the nations assembled to create one insect, they [could build its body but] could not endow it with life."[7] Before Creation, by divine will, the Infinite Light[l] filled all "space" (meaning spiritual space). Finite existence was not possible, and because physical existence is finite, there was nothing physical. Finite light was, as it were, absorbed and concealed within the Infinite Light in the way that the light of a candle is nullified in the brilliance of the sun.

When God wished to create the universe, He withdrew His Infinite Light and concealed it within Himself, allowing the revelation of His finite power and light and making room for finite existence. This is called the first tzimtzum, the Hebrew word meaning concealment and contraction (plural tzimtzumim). It was not a withdrawal of God Himself but only the concealment of His Infinite Light. Tzimtzum involved no change in His essence or omnipresence; it was merely the removal of the manifestation of His Infinite Light from actuality into latency. This first contraction was therefore different from all subsequent ones because it revealed, for the first time, something finite. All subsequent tzimtzumim were further concealments, but from finite to finite.

Yet even this finite power, the source of all Creation, was too abundant and powerful to give rise to finite beings. Only after an innumerable number of tzimtzumim, which progressively condensed and concealed the creative power, was it possible for finite and physical

---

[k] Genesis describes only three out of nothing creations: the beginning, animal souls, and human souls—or whatever you like to call what vivifies animals and us.

[l] When discussing the Creation process, Kabbalists equate God with the sun (a luminary that emits light), and His emanations are likened to light from the sun. God is referred to as the infinite one or *Ein Sof*, and his emanating light is called light from the infinite, the *Ohr Ein Sof* (in Hebrew, *ohr* means light). God the luminary never changes. His light emanates in all directions and fills all space. The light has components, analogous to the colors of the rainbow. His light can be contracted and filtered, as sunglasses act upon sunlight. The light can go into containers and emit a particular component, as is the case when we shine light into a red vase and see only red light from the outside of the vase. For a full explanation of the analogy, see Appendix B: The Light Metaphor.

matter to exist. Through this concealment of the power by which all things exist, the world appears to enjoy an independent existence, as if it were apart from God.

The tzimtzum process led to the point where the initial physical space appeared. As the process continued, this space was filled with the elementary particles and the laws of physics—all the ingredients to build a universe, as in our Lego analogy. In fact, the Hebrew word chalal, used to describe this primordial space, has the same meaning as "events." In other words, the space was full of events, just as physicists describe the original quantum vacuum: full of short-lived particles coming in and out of existence.[8]

Let's review the order of events leading up to the point where all the ingredients required to make the universe have appeared.[9] Using a table helps us match the description found in Genesis to the full account of Creation, as elaborated in the mystical sources, to the physical manifestation observable with scientific instruments. Most of the terms will be familiar from Part 1; a couple will be new and are elaborated in footnotes.

Chapter 14 – Before the Big Bang

Creation Process: Biblical		
Genesis	**Full creation account**	**Physical manifestation**
	God and the blueprint exist. No space, no time. Nothing.	
	Spiritual time exists, allowing for a process in an order.	There exists an order of time[m,10] (which includes a direction, the "arrow of time").
In the beginning... Gen. 1.1	Tzimtzum. God contracts His light in a central point and leaves an evenly measured place on all sides: the chalal or void. It is expanding.	Space, finite and expanding, exists. The size of space at the beginning is finite (a few mm).
In the beginning... Gen. 1.1	Reshimu.[n] He leaves behind a residue in the void, which fills the void with potential.	Physical entities can exist in space. Time exists as the measured motion of physical entities/particles.
In the beginning... Gen. 1.1	Kav.[o] The creation of the finite world begins.	

---

[m] The order of time is God's plan (or sequence of events) from beginning to end, which He sees in one glance. It is not physical time; think of it as a to-do list.

[n] The reshimu is the residual impression of the infinite divine light that God "withdrew" from Creation through the process of tzimtzum. The reshimu is sufficiently "weak" and virtually "invisible" ("nonexistent") to allow for the existence of independent reality and to serve as its divine "background."

[o] To help us visualize what happened after the initial tzimtzum, Rabbi Chaim Vital suggests imagining a circle that is full of the Ohr Ein Sof. No finite existence can be created within this circle because the Light of the Infinite totally occludes it. The first tzimtzum concealed the Ohr Ein Sof so that within the circle was left a void within which something finite could be created. In the next stage of Creation, a beam of pre-tzimtzum light called the kav was introduced into this circle. Contained within this light were all the ingredients for the formation of the various worlds.
https://www.chabad.org/library/article_cdo/aid/361884/jewish/Tzimtzum.htm

	Creation Process: Biblical	
Genesis	**Full creation account**	**Physical manifestation**
...*was (tohu vabohu) chaos and desolation* Gen. 1.2	Tohu. The world of chaos is created. The light from God enters vessels that are vertically arranged and independent and cannot contain the light.	Space is like a vacuum with energy in chaos. It is a soup of matter and antimatter coming into and going out of existence.
...*was (tohu vabohu) chaos and desolation* Gen. 1.2	The breaking of the vessels[p] of tohu leads to a massive release of energy. The remains of the vessels' shattered fragments are the basic physical constituents of the world.	The breaking releases a large amount of energy, leading to the further expansion of space. Various symmetry breaks occur, including the separation of the forces.
		It also creates all the elementary particles. These are not restrained by space–time and thus obey the physics of the small: quantum mechanics.

---

[p] The breaking of the vessels of tohu was explained in Chapter 9. A good way to visualize why it happened, and why at this point, is with the analogy of an architect who wants to build a palace. First, he arranges for all of the materials to be at hand, then the actual construction begins. Similarly, God prepared all the materials He would require. However, by the end of tohu, they were in a confused heap. He then proceeded (via tikkun) to put them in order and assembled them into structures (via speech). (From Rabbi Chaim ben Attar, translated by Eliyahu Munk, *Or Hachayim: Commentary on the Torah* [New York: Lambda Publishers, 1999], Genesis 1.1.)

Chapter 14 – Before the Big Bang

	Creation Process: Biblical	
Genesis	**Full creation account**	**Physical manifestation**
God said, "Let there be light"... Gen. 1.3	Tikkun. The world of rectification is created, where the light from God enters new vessels that can contain it. Tikkun and tohu are distinct and separate.	Macroscopic entities exist only in vessels, so they are constrained by space–time and obey the physics of the macroscopic: general relativity. The macroscopic is fundamentally different from the microscopic.[11]
God said, "Let there be light"... Gen. 1.3	Speech. Everything is made from combinations of the 22 foundation letters of the Hebrew alphabet.	The universe develops through the various statements in Genesis. Everything is made of twenty-two elementary constituents: 12 matter particles; 3 force carriers; and 7 mass and gravity-related particles.
	Everything in Creation exists because of a life force vested in it, which descends and is progressively diminished by transpositions of the letters and their numerical values.[12]	These elementary particles have unchanging properties because they contain a life force or spark. Those at the lowest level (which make the physical world) are eternal. The particles' properties can be inferred from the letters' shapes, sounds, and numerical values.

167

So, if the Torah tells us what came before the Big Bang, why do we keep searching for a natural explanation? Apart from the beginning of the physical world, everything else has an apparent link with a source of existence. For example, light comes from a luminary, speech comes from a thought. "When viewing material matter, however, at the essence [i.e., the beginning of the universe], one does not perceive that it derives from ... something higher than itself; it seems to exist [as] a wholly autonomous thing."[13] Why?

Because the beginning was created out of nothing, so we are incapable of comprehending its source: God.[14] Hence, we call the source "nothing," meaning that the source does not exist within our limited world and does not exist in the same manner (i.e., within our range of comprehension) as we do.[15]

Chapter 15

Time and Relativity

Rhythm: A sequence of regularly recurring functions or events.

We return now to our questions about time. What is it? How did it start? Why does it flow in one direction? And what is its purpose?

The Science Answer

The original picture of time that fits with our intuition is that time flows uniformly and equally throughout the universe, in the course of which all things happen. In this picture, the past is fixed and has occurred. The future is open. Reality flows from the past through the present into the future.

However, modern science has discovered a very different and more confusing picture, which leaves the questions above largely unanswered. Despite this we are totally at ease working with time in science. We take a pragmatic scientific approach by using a definition of time that works with all our equations and that allows us to calculate what happened in the past and will happen in the future. A deep understanding of time has been superfluous because Newton's laws, Einstein's theories, and quantum mechanics don't require us to understand the nature of time to do the calculations.

We often talk about measuring time—when we clock how long it takes us to jog a mile, for instance. But do we actually *measure* time? When we determine temperature with a thermometer or distance with an odometer, we interact with the thing being measured; energy is exchanged, and we get a measurement. When we "measure" time, though, we don't interact with it at all. No one has felt time in the way we feel temperature. What we are actually doing is counting the cycles of something else (a pendulum, electrons changing energy levels in an atomic clock, the rotation of the Earth), not measuring time.

We've also discovered that the number of cycles we count is affected by the speed at which we're traveling and the local force of gravity. In 2015, the speed effect on time was observed even at a typical bicycling velocity; this was determined with the latest atomic clocks, which are accurate to one second in fifteen billion years.[1] In terms of gravity, different places on Earth with different elevations yield different cycle counts. So, all we actually count are the cycles of something, which is not necessarily the same as time.

In short, physics has not told us what time is. Furthermore, although many believe time came into existence with the Big Bang, some—in particular, those who think this bang was just one of many cycles of the universe's expansion and contraction—think that time preceded "our" Big Bang and perhaps has existed forever.

If Einstein was right that time is relative, depending on where you are and how fast you are moving, so there is no universal time, how do we come up with the age of the universe? We postulate a definition of time to use with the Big Bang theory: cosmic time. We make some reasonable assumptions about the expanding universe,[2] and then we measure the passage of time using imaginary clocks that move with the flow of the expanding universe rather than being in some arbitrary place, such as Earth. We then define the beginning of time, or time zero, to be when the Big Bang happened, and we count forward. Not clear what this means? No worries—it's just a definition that allows us to calculate the age of the universe and formulate equations for the expanding universe. The point still stands: we can work with time, but we still don't know what it is.

Why does time flow in one direction? As we mentioned in Part 1, there are theories, but no definitive explanation. Physical processes at the microscopic level are believed to be either entirely or mostly time-symmetric: if the direction of time were to reverse, the theoretical statements that describe them would remain true. Yet at the macroscopic level it often appears that this is not the case: there is an obvious direction (or flow) of time. The best candidate to explain the directionality of time from past to future is the fact that many processes are irreversible. In other words, eggs break and scramble but they never go from broken and scrambled to whole. What provides this directionality or arrow of time is the second law of thermodynamics, which says that in an isolated system, entropy tends to increase with time. Entropy can be thought of as a measure of disorder; thus this law implies that as a system advances through time, it becomes more statistically disordered. It is this asymmetry that we think leads us to distinguish between future and past. But is this perception reality? Physics does not yet know. And if it's real, who made the original world so ordered that we could forever disorder it? Or do we live in a special place that's ordered, whereas the rest is disordered—so only here, by accident, do we perceive time?[3]

More philosophically, we could ask, "What is the purpose of time?" Physicist John Wheeler provided the following workable answer: "Time is what prevents everything from happening at once."

The Biblical Answer

As we saw in Part 1, the Torah provides quite a detailed answer to the "What is time?" question. Here, we'll recap the earlier discussion and then go deeper into the nature of time and its flow. Then we'll explore further afield, including what the purpose of time is.

Let's start at the beginning with how time—*asman*, in Hebrew—came to be.

Asman and Seder Asman

Prior to creating the physical universe, God created the entirety of time in a single instant. What we think of as physical time is the unfolding of that instant into myriads of particulars, which we then experience as a succession of moments aligned one after the other.[4]

God is above time.[5] Let's review His "process" of creating time. He first brings into existence an "order of time"—*seder asman*, in Hebrew—in other words, His plan from beginning to end, which he sees in one glance. This is not physical time as we know it but a spiritual precursor. I find it easiest to think of the order of time as a numbered to-do list—an ordered list of things that have to be done, but with no assigned schedule and thus no way to measure progress.

This sequence of events will determine the one-way arrow of time that we eventually observe in the physical universe. However, in this "one glance" spiritual state, all of the events are visible at any one moment. To help you grasp this, imagine the way you think about items in a space—say, in your living room. The furniture is arranged spatially, and you can look left and right and see the whole thing in "one glance." In a similar sense, while physical time is chronological, spiritual time is not; it's like objects in a space.

Consider another example, this one from mathematics. We say 1+1=2, and therefore 2−1=1. These statements follow one another chronologically in the sense that the first leads to the second. However, the fact that the second equation follows the first doesn't mean there was a point when one existed and the second did not. In abstract form, they both have always existed—we just find it easier to comprehend them by assigning a chronology to them.[6]

At some point, God withdrew His Infinite Light, leaving a small residue that allowed for Creation. This residue, the "void,"[7] eventually devolved to physical space, which contained a very thin substance (tohu) from which everything was made.[8] Once physical creation exists, with physical objects that move (such as particles), there is physical time as we experience it; in the biblical sources, this physical time is described as measured motion—in other words, the counting of the cycles of something physically moving.[9]

However, time has a cosmic organization derived from its spiritual predecessor, the order of time. Time came into being through a series of contractions. The "one glance" was broken down into detail as it evolved from its spiritual form to the physical form. Time was first split into the days of Creation and the six millennia of history (each millennium corresponding to one Creation day).[10] Each millennium was then split into years—which include the entire 365 days in one glance—and these were split into months and the months into days. The days were then split into twenty-four periods:[11] twelve night hours and twelve day hours. Each of these hours corresponds to a permutation of one of the names of God—the essential name, or YHVH, with its twelve permutations[12] for the twelve hours of night, and the name Adonai ("Lord" in English) with its twelve permutations for the hours of day. Each of these includes all the minutes of one hour in one glance.[13] The breakdown continues to smaller and smaller units.

Let's envision this process of breakdown as a pyramid. At the highest point of the pyramid—representing the divine act of creation—all of time and history is encapsulated within a single point. As the creation process ensues until the physical creation manifests, we proceed down the pyramid, and each slice through the pyramid represents a further expansion and division of time. High up the pyramid, slicing produces a small base that represents the split into the Creation days and the millennia of history. At the bottom of the pyramid, we have a wide base, representing the physical time we experience down to fractions of a second.

There is no limit to the divisions, so time is continuous, not quantized like many other creations (particles, for example). So, the only way we can experience or at least "measure" physical time is by perceiving the movement of physical creations in space. If nothing ever changed or moved, we would not experience time. This is the time we know and live within—"measured movement"—and it is, by definition, relative, dependent on whatever affects the motion, particularly the speed of travel and the force of gravity.

However, for practical everyday purposes, there is a universal time, which is what we measure on our planet. It is based on the rotation of the Earth around its axis and the sun, and the rotation of the moon around the Earth:[14]

> and God said "let there be luminaries in the expanse of the heavens, to separate between the day and between the night and they shall be for signs and for the appointed seasons and for days and years."[15]

The Arrow of Time

What we experience as the one-way arrow of time through the verb tenses of past, present, and future is just the last, most concrete incarnation of what starts as the order of time, descends through the many contractions described above, and is broken down into details (millennia, years, seconds…). Because it is only experienced as the "measured motion" of things, it appears to be causative and to flow in a particular direction.[16]

Although our experience is that time flows—we can stand still, but time still marches by—in the biblical conception of time, it forms the background or terrain in which we exist, and this terrain is part of a whole plan or map for history. The Torah refers to the holidays as mo'adim,[17] usually translated as "appointed holidays." While one meaning of mo'adim is appointment in the sense of meeting—festivals are meetings between the people and God—it also means a specific period of time that possesses an overarching characteristic and serves a purpose. It's more like a terrain, with each feature having a purpose. In this sense, time does not flow but forms a background on which we exist. Its "terrain" is causative in that a particular holiday—say, the remembrance of the Exodus—is not an anniversary as we are used to thinking, but is the time at which a particular characteristic is accessible. In the case of the time when the Exodus is remembered, freedom is the accessible characteristic; that is why the Exodus happened then.

Time is also like an organism, "a whole with interdependent parts." This means that each period of time belongs to a complete plan of history; it is all one "organism." In the same way that instructions flow from our brain to the rest of our body, not vice versa, we experience the arrow of time running from past to present.

Finally, time is designed with distinct qualities, one for each day of Creation (and therefore for each millennia of history)[18] in which a dimension of time was brought into being. In other words, not only is time per se an original creation, but the dimensions and cycles by which it's measured and defined are also entities created by God.[19]

These characteristics or revelations of time that correspond to the days of Creation also manifest in the parallel six millennia of history, which are broken down into years, months, and the days of the week. Each of these millennia, months, days, etc. has a particular spiritual quality that affects what manifests and happens then (subject to our free will, of course). Thus, Sunday is the best day for loving kindness, Monday is the best day for discipline, and so on.[20] We won't go into the details of the tapestry of time and its impact on our lives,[21] as these aren't relevant for understanding what time is.

Figure 15.1 Time manifests from the characteristics of the six days of Creation to the details of a fraction of a second

The Role of Time

So why did God go through all this trouble? What's the purpose of time?

We saw in Chapter 10 that biblically, there are five dimensions:[22]
- "world," which includes the three spatial dimensions
- "year," the time dimension
- "soul," our ability, due to free will, to choose between right and wrong at every point in time and space

The sages explain that there is a very large tension between the world and soul dimensions.[23] In the world dimension (the realm of tikkun or classical physics), if something is here, it is not anywhere else, and whatever is far away cannot be here. The physical limitations of the world are *he'elem*—an envelope covering and concealing the infinite divine light. Both physically and spiritually, we become fixated on the fact that it is impossible to be in two different places and two different situations at one and the same time.

The soul dimension (our spirit) refuses to accept these limitations. It is strongly attracted to the infinite, to the source of everything, without limits, without a world that is defined as a place, and without any fixity in feelings or emotions. The soul is the point of light and spirituality above any details of the physical world. It wants to violate the constraints of the world dimension and be in two places at once.

This is why time needs to exist. Time is in the middle, a mediator, serving as a link between the soul and space dimensions. Time allows us to go from one place to another, no matter how far, thus forming a bridge between the soul's limitless view and the constraints of the space dimension.

Similarly, time allows us to comprehend what goes on around us and to function. If we did see everything in one glance, we would be totally confused. Our thoughts seem infinite, always on the move, and can contain contradictions, but our actions are limited to a specific time and place. We can only process and understand things in some kind of sequence provided by the march of time. The role of time is thus to allow speech, which in turn "translates" thought into the practical world of action. Thought is infinite and contains everything at once, whereas speech takes time and is sequential, so it translates thought into a sequence of words, one word at a time, which can then be turned into one action at a time. In this way, the infinity of thought is decomposed into a time sequence—expressed as speech—for the world of action, where one thing happens at a time, in a place.

In fact, that is why God created the world in six days. He could have done it in an instant, but we would never have understood how the universe came to be. Instead, he spread out the universe's creation over time—speaking it into existence sequentially—so we could understand it and even develop evolutionary theories to explain it!

Timeless Light

We've seen that all creations devolve from spiritual counterparts. In particular, light has as its counterpart spiritual light, the Ohr Ein

Sof.[24] Thus, we can say that the physical creation—light or electromagnetic radiation—derives from the spiritual root, the Ohr Ein Sof, where there is no time because God is above time. Thus, a ray of light does not change and experience time. This leads directly to the essence of special relativity: the speed of light is fixed, and at that speed, time does not elapse.[25]

Chapter 16

When Microscopic Meets Macroscopic

Let's return to the question of what happens when the microscopic world meets the macroscopic world. In quantum mechanics, this is known as the measurement problem. In other words, what happens when we use our large measuring equipment, which is governed by the laws of classical physics, to measure something about a very small particle, such as the position of an electron, which is governed by quantum mechanics?

The Science Answer

The quantum world does not seem to mix with our macroscopic world. With quantum mechanics, we can make incredibly accurate predictions about the probabilities of events—for example, where we can find a particle after it has gone through a particular opening. But we can only actually see or confirm these predictions by measuring the position of such a particle over and over again until we are able to calculate a probability of it being at any point and thus verify the prediction.

In quantum mechanics, the quantum state of something is given by a wave function: a mathematical description of the state of something—an electron, say—that can be used to calculate the probability of some property of it, such as the probability of the electron being at a particular place. We don't actually know where the something is until we make a measurement. At the point of making the measurement, we say that the wave function collapses—meaning that instead of describing a number of possible outcomes (and providing the probability that the particle is at any one of them), it collapses to one answer, the measured position of the particle. We don't understand what happens when we measure the position of a particle. How does the wave function collapse to provide an actual position?

Chapter 16 – When Microscopic Meets Macroscopic

We don't know. Nor do we really know the status of the particle prior to the measurement. Was it in many places at once, and when we measured, we fixed it according to the predicted probability? Or was it in a well-defined place, but we just didn't have the information about exactly where until we made the measurement? Or…?

Thus, although quantum mechanics has withstood rigorous and thorough experimental testing, many of these experiments are open to different interpretations of what is actually going on.[1]

The standard interpretation of quantum mechanics, known as the Copenhagen interpretation, takes a very pragmatic and simple approach to allow calculations. The question of where the particle was before its position was measured is not asked. Only the act of making a measurement with a large, classical sensor is considered, and this gives rise to the particle being found in that particular location. Thus, the equations are just used to calculate results, without the need to understand what's going on.

The main question in this interpretation is: What constitutes an actual measurement? Some who use this interpretation believe a conscious human has to be involved in the measurement. How this causes the wave function to collapse is widely disputed. The problem of whether consciousness is required to make the measurement worsens when we consider the Big Bang, an event where the universe was small enough for quantum mechanics to apply, yet no human was there to measure anything.

In the "many worlds" interpretation of quantum mechanics, the wave function doesn't collapse, so the problem is eliminated. However, this interpretation implies that all possible alternate histories and futures of the particle are real, each representing an actual "world" (or universe). When we measure the particle, we just happen to observe one particular outcome in a particular universe—the one we are also in. Thus, there is a very large—perhaps infinite—number of universes, and everything that could possibly have happened in our past, but did not, has occurred in the past of some other universe or universes. However, the many worlds theory creates almost as many philosophical problems as it solves.[2] If you find the theory a bit confusing, don't worry—most people do.

In the "hidden variables" interpretation, the position of the particle is determined, and hidden variables indicate what that position is—we just don't have access to them. We only have access to the wave function, which provides some information but not all, making reality seem probabilistic. However, decades of rigorous experiments (that nonetheless remain controversial) seem to show that some of the hidden things would have to operate faster than light, breaking perhaps the most fundamental law of physics.[3]

None of these interpretations can be totally confirmed, and all have issues. Other interpretations continue to be worked on. In the 100 years or so that quantum mechanics has existed, we've been able to work with it but not understand it. Today, many physicists continue to work on an interpretation of quantum mechanics that is a variant of the Copenhagen interpretation, one that specifically solves the need for an observer to make the measurement. This idea, still under development, is known as the decoherence theory. Decoherence in essence says that isolated microscopic particles do behave quantum mechanically, but in practice, when those particles interact with the large environment—such as when a measurement is made—the interaction basically stops the quantum behavior and forces the particle to exhibit a particular set of characteristics—for example, to occupy a position. Under this interpretation, the particle simply needs to interact with the macroscopic world; no observer, human or otherwise, is required.

The Biblical Answer

Let's review the relationship between God, time, and space.[4] The divine transcends time and space. They are completely nullified in relation to God's essence and being, just as sunlight while still within the sun is nullified there. Thus, the dimensions of time and space have no relation to the holy supernal attributes of the highest world, because those attributes are infinite. Nonetheless, although God transcends time and space, He is also found below, within time and space.

The Torah recognizes many creations that are beyond time and space. We don't need to explore intangibles, such as the higher worlds, to see this; there are everyday creations that transcend time and space.

Chapter 16 – When Microscopic Meets Macroscopic

Mathematics is true, yet much of it (excluding geometry) has no relation to space or time. We saw earlier that we may assign a chronology to mathematics to make it understandable—for example, by saying 1+1=2, therefore 2–1=1—but both statements are true at all times and in all places.

It's therefore not surprising that the Torah recognizes a physical world not completely bound by time and space—tohu—and one constrained by time and space: tikkun.

The Worlds of Tohu and Tikkun

Fundamentally, the Torah says that we have two existences: one from the world of tikkun, and one left over from the world of tohu.[5] In particular, general relativity, classical physics, and macroscopic things belong to the world of tikkun,[6] quantum mechanics and small particles to the world of tohu. These two worlds are different, so we should not be surprised by the apparent clash between general relativity and quantum mechanics.

As we have seen, the world of tikkun is our everyday world. The Godly energy that makes things exist is enclosed in vessels, so the material things in this world are constrained by time and space. However, the world of tohu is different. The Godly energy is not contained in vessels, because the vessels broke. Let's review what happened, as hinted at in Genesis 1:2.[7]

The existence of the finite world as we know it, and as God intended it, was not possible until the Godly light went through several stages of contraction and descent. In the early stages, the light was put into the vessels of tohu. Ten individual vessels were formed for the light. They were emanated in such a way as to be situated one above the other, in a single line. (This was unlike the vessels in the world of tikkun, which were arranged in harmonious intersecting triads.) Thus, each vessel of tohu existed as an autonomous fiefdom, so to speak, independent of and even in opposition to the others. As the light was released into the vessels, the first three held and contained the light. However, when the fourth vessel received the light, it broke, and so did all subsequent vessels.

As the parts of the vessels were projected downward through more contractions, they broke further into ever-increasing numbers of fragments, and the residual bits of light attached to each fragment also fragmented further. As these fragments continued falling, they became not only more numerous but also more tenuous. Because these fragments retained only a spark of the full light to sustain their existence outside a vessel, they formed the root elements of all creation—in scientific language, they became the elementary particles. As these particles are not in vessels, like macroscopic objects, they are not constrained by time and space, so they can be in two places at once and affect their past, as quantum mechanics experiments have "observed."

After the vessels broke, the process continued with the light going into the orderly interconnected vessels of tikkun that did not break. These in turn eventually devolved into the macroscopic world. So in this world of macroscopic creations (basically, the world as we see it), everything is in vessels, bound by time and space, located at one point and time only.[8]

What happens when you take something unconstrained by time and space and try to constrain it? When something from tohu is forced to be confined in a small space, it "goes mad," which is what quantum particles do when confined,[9] i.e., you can't tell both their position and speed precisely at the same time, so if you precisely fix the position, the speed becomes undefined. However, this does not mean the two worlds can't interact.

How Do These Worlds Interact?

How can tohu and tikkun be unified? According to the Torah, this can occur by inter-inclusion, but I haven't yet been able to work out how that would appear to us in the world of physics.[10] We do know, though, how the two worlds interact. The Torah says that quantum mechanics and classical mechanics are by nature separate realms. The microscopic particles described by quantum mechanics are not constrained by time and space, whereas the macroscopic objects in our everyday life are. Yet while microscopic objects can exhibit spooky behavior because they're unconstrained, the moment they interact with anything macroscopic—such as a measurement device—they become constrained and have known, definite outcomes.

We cannot "see" what's going on in the realm of tohu, so we don't know what happens to quantum particles before they interact with the macroscopic. But when tikkun observes or interacts with tohu, it puts the tohu piece into a vessel and constrains it in time and space. Whatever vessel it puts it into completely affects it. This means that when we measure the micro realm from the macro realm, we amplify the micro so we can record it, because every experiment is an interaction between the apparatus and the particle.

In that amplification process, what we observe depends on what we do in the experiment. If we amplify wave behavior—meaning that we catch the tohu existence via an experiment that puts it into a wave-like tikkun vessel—we see a wave. If we amplify particle behavior, we observe a particle.[11] If we want to measure position, we amplify that and get a position for the particle (but not its momentum), and so on.[12] Note that no measurement or observer is required, only an interaction between the macroscopic tikkun and the microscopic tohu.

In other words, the Torah asserts the Copenhagen interpretation, which says that we cannot ascribe a definite meaning to a quantum event until a measurement is made and the result is amplified to the macroscopic realm—tohu is put in a tikkun vessel. However, the microscopic *can* go into a vessel without an observer making a measurement.[13] For example, with no humans around, a photon from the sun can hit the surface of a table and turn to heat. Given that the Torah says no measurement of the tohu particle is required for the particle to become part of the macroscopic, it offers a similar interpretation to the main one being worked on by science today: decoherence.

A crude analogy may help clarify the Torah interpretation. Let's assume the quantum particles are analogous to water drops. These behave in very particular ways: they are spherical, stick to car windows sometimes without sliding down, form dew drops on blades of grass, etc. If we are to describe the physics of these drops and their behavior, we need special equations that take into account the surface tension of water. Now assume that the macroscopic world is analogous to a large body of water, such as a lake or a pond. The equations of physics for these appear to be different from those used for water drops. What happens when a water drop meets a lake—in other words, is absorbed into a large world of many drops? It loses its identity and behaves totally differently, like water in a lake. It's no longer discernible as a sphere, doesn't stick to vertical windows, doesn't form drops on blades of grass. It behaves like one big mass of water. The drop of water can meet the lake naturally or by a human picking it up with an eye dropper and depositing it in the lake. The human doing this knows the position of the drop, its weight, and whatever else about the water drop he wants and is able to measure. But in the end, either way, the drop joins the lake. The quantum is absorbed in the macroscopic—no need for an observer. Similarly, if a land slide hits the lake, water is splashed in the form of drops, which once again obey the physics of drops. So the two apparently different worlds go back and forth with or without the intervention of an observer, human or otherwise.

String Theory?

One proposed solution in physics to the issue of quantum mechanics and general relativity coexisting is that the fundamental components of the universe are not particles—infinitely small point sources—but strings. These strings are the most fundamental entities and are finite in size, and their vibrations emanate particles. So, it makes no sense to discuss what happens when dimensions are smaller than the strings. In effect, string theory bypasses the whole issue of quantum mechanics and general relativity coexisting by not allowing anything to get smaller than a string.[14] String theory predicts certain heavy particles that so far have not been found.

Whether string theory is correct or not, note that the biblical answer effectively asserts the same solution, because it says that the smallest entities are manifestations of the Hebrew letters. These letters, except the one corresponding to the electron, have an extended shape and thus devolve into physical particles that are finite in size, and their most elementary property is sound or vibration because they devolve from letters, which are "spoken."

Chapter 16 – When Microscopic Meets Macroscopic

The Big Bang and Black Holes

The key area where the quantum and classical world meet is where the density is great, the forces are great, and the space is tiny. We've seen that if the Big Bang started from an infinitesimal space containing the mass and energy of the visible universe, then we need both realms to work together for us to compute what went on. However, the Torah says this did not happen. The original piece of space was not infinitesimal, it was small but macroscopic, and the two worlds did not need to coexist.

What about black holes? Our general relativity equations say that as a star implodes under its own weight and forms a black hole, it will keep doing so until it's an infinitesimal point. In that confined space, we then need quantum mechanics to work with general relativity to understand what's going on. But what if this is incorrect? After all, at this point the equations become undefined so we don't know what actually happens. What if something happens as the center of the black hole, containing its mass, gets smaller and smaller, and it never reaches an infinitely small point? After all, this seems more reasonable. In our water drop analogy, this would be equivalent to us putting water under pressure and squeezing it through a very small opening, like when we block most of an opening of a hose with our thumb. What happens? The water—macroscopic object—turns into droplets. It doesn't squeeze into nothing; it returns to its microscopic form. Maybe the same happens as matter is compressed in a black hole: general relativity gracefully gives way to quantum mechanics, and the matter turns into interacting particles similar to the original quantum vacuum, in a finite, albeit very small space.

Chapter 17

Elementary Particles

We return now to elementary particles. What are they? How many are there? And why do they have the properties we have measured?

The Science Answer

In Chapter 6 and Figure 6.2, we described the Standard Model of particle physics: the theory describing three of the four known fundamental forces in the universe (the fourth being gravity) and classifying all known elementary particles. We saw that as powerful and as accurate as the model is, it leaves many things unexplained and unanswered. In the words of Brian Greene:

> The standard model ... makes impressive accurate predictions for how the particles will interact and influence each other. But the model can't explain the input—the particles and their properties ... the detailed features of the elementary particles are entwined with what many view as the deepest question in all science: why do the elementary particles have just the right properties to allow ... life to exist?[1]

The model has three generations of particles, but why three? We have no explanation.[q] The particles have several fine-tuned properties, all measured but not explained. The forces have very specific strengths. Why do those and only those work for our universe? The model is incomplete because it does not deal with gravity and dark matter. Finally, it does not explain why our universe has a large excess of particles instead of an approximately equal mixture of particles and

---

q For a full description of the particles and properties, see Appendix A.

antiparticles. This is known as the matter–antimatter asymmetry problem.[2]

Several "beyond the Standard Model" theories have been proposed (such as string theory), but so far, experiments are still too consistent with the Standard Model to provide guidance on how to enhance it.

The Biblical Answer

In Chapter 12 and Figure 12.1, we saw the biblical standard model, depicting the twenty-two Hebrew letters that were used to make the universe. We now return to a more in-depth look at speech and the Hebrew letters.

God's Speech

In the Kabbalistic writings, how is God's speech related to creation? Scripture states, "For He spoke—and it came into being,"[3] meaning that Creation resulted from God's speech and the breath of His mouth.[4]

How exactly did this happen? He contracted the light and life force that could diffuse from the breath of His mouth and invested it in the combinations of the Hebrew letters that made up the Ten Sayings of Creation. The Ten Sayings are the ten creative statements in Genesis, such as "Let there be light." Everything else not mentioned in the Ten Sayings was then made by: (1) combinations of the letters—that is, words; or (2) substitutions and transpositions of the letters themselves—that is, by rearranging letters according to various letter transposition systems in Kabbalah;[5] or (3) their numerical values and equivalents, meaning that the numerical value of the letters also brought things into being. There are several ways to assign numerical values to the letters.[6] This, as we've seen before, was a spiritual process that devolved to a physical creation after many contractions.

Why go through this whole process? The original letters are used for the important creations in the Ten Sayings. Each substitution and transposition causes a descent of the light and life force degree by degree, so that it creates and gives life to creatures whose quality and significance is lower than the quality and significance of the creatures created from the actual letters and words of the Ten Sayings. Similarly, progressing down to the numerical value yields the progressive diminution of the light and life force all the way to the point that the life force can invest itself even in the lower created things, such as inanimate stones and dust.

> In summary: The Divine life-force is capable of creating worlds that are infinite both in quantity and in quality. Finite beings are created only when this life-force garbs itself in the letters and transpositions of the letters of the Ten Utterances [i.e., the Ten Sayings] and in their numerical values.[7]

Organization of the Letters

In Chapter 12 and Figure 12.1 we encountered the Kabbalistic diagram depicting the arrangements of the letters into three levels and three columns. We then learned about the correspondence of the letters to particles in the Standard Model, and we determined which letters related to undetected dark matter and gravity, based on their positions in the diagram.

Information in the Letters

We also explored how each letter's shape provides information. What about the rest of the information that science has found but cannot correlate? For example, we've measured the mass of each particle but have no idea why it is that particular value or how, if in any way, all the masses of the particles are related. It's clear from the above discussion on God's speech that the information must be encoded in the letters. I have taken an approach to try to deduce the mass of all the particles from the numerical value of the letters, but this is work in progress. I've yet to deduce other parameters, such as spin.

There are 613 commandments in the Torah for the Jewish people. These are divided into 248 positive commandments—things we are to do to build the world into a better place, such as charity—and 365

Chapter 17 – Elementary Particles

negative commandments, things we are not to do so as not to make the world into a worse place, such as stealing.

The 248 are associated with the right-hand column in Figure 12.1, the 365 with the left-hand column,[8] and the 613 with the middle column. By assigning these values to each of the three columns and then using the numerical values of each letter, I have produced the following table as a work in progress.[9]

Column 1 of Figure 17.1 contains the particle names (where I have provided a descriptive name for those particles still to be discovered). Columns 2 and 3 contain the corresponding Hebrew letters and their numerical values. Column 4 contains the scale factor derived from the position of the letters in Figure 12.1. Column 5 contains the particle masses predicted from the Hebrew letters (using the measured electron mass and the scale factor). The actual measured mass for each particle is presented in Column 6. These mass values are obtained from the Particle Data Group 2020 update.[10] Column 7 shows whether the predicted masses and measured masses are within the measurement error range. Predicted masses closely match actual measured masses, except for the bottom quark.[11]

Most of the particles' masses have been derived from the letters' numerical values and position, within experimental error. Similarly, the masses of the predicted four dark matter particles have also been derived, as well as the mass of another Higgs-like particle. I don't know whether these predicted mass values are correct or not, since there is only one discovered particle (the Higgs) in the whole group of particles in the vertical connectors of Figure 12.1, but they illustrate that the letters contain a great deal of information, as explained in the earlier section on God's speech. One can also compute an energy for the massless particles like the photon, which comes within measurement error of the Planck energy,[12] theorized to be the largest energy a photon can carry. The Planck energy is calculated by using three fundamental constants of nature, whereas in this model its derived from the mass of an electron!

1	2	3	4	5	6	7
Particle	Letter	Letter value	Scale factor	Predicted mass (MeV/c2)	Measured mass (MeV/c2)	Error range
up quark	lamed	30	0.075	2.3	2.2	within range
down quark	samesh	60	0.075	4.5	4.7	within range
charm quark	tzadi	90	13.725	1235	1270	within range
strange quark	zayin	7	13.725	96	93	within range
top quark	ayin	70	2505	175342	172760	within range
bottom quark	vav	6	2505	15029	4180	out of range
electron	yod	10	0.051	0.51	0.51	control
muon	kuf	100	1.056	105.6	105.7	out of range
tau	hei	5	355.441	1777.2	1776.9	out of range
dark 1	pe	80	6.90	552	undiscovered	
dark 2	gimel	3	1259.3	3778	undiscovered	
dark 3	kaf	20	0.554	11	undiscovered	
dark 4	bet	2	178.2	356	undiscovered	
Higgs boson	resh	200	626	125297	125100	within range
Higgs-like	dalet	4	192018	768070	undiscovered	
Z boson	alef-vav			91188	91188	control
W boson	alef-yod		0.883	80525	80379	out of range

Figure 17.1 Predicted particle masses corresponding to Hebrew letters

Chapter 18

A Theory of Everything?

Why does science succeed by looking for simple, unifying concepts—ultimately, a theory of everything? And why are scientists drawn to this method?

Advances in science have often been driven by the search for very simple, unifying concepts. Early on, a Greek philosopher conceived of all matter as being made up of tiny, indivisible particles, which he called atoms. Subsequent generations of inquiring minds continued the process, and although the Standard Model is not as simple as originally envisioned, it's still straightforward compared to the world. Newton looked to link what made an apple fall to the ground and how the moon was held by the Earth, arriving at his "simple" theory of gravity—so unifying that many scientists thought there was little else to work on. Einstein made gravity even more powerful and unified by showing that mass tells space how to curve, and space tells matter how to move. Schrödinger encapsulated quantum mechanics in a simple, powerful wave equation.

Scientists have long believed that the four forces of nature can be unified as one force, which existed early in the development of the universe and then split stepwise into four during the first fractions of a second after the Big Bang. Two of the forces have been unified, the third is almost there, but gravity still eludes unification—yet most scientists believe it won't for much longer. They also believe we will unify Einstein's theory of gravity with quantum mechanics.

We've further discovered that there are multiple valid ways of describing so many physical phenomena—all making equivalent predictions, yet all wildly different in their premises. But an even stranger fact is that when there are competing descriptions, one often turns out to be truer than the others because it extends to a deeper or more general description of reality. We saw an example of this in quantum mechanics, although we don't yet know whether all of its interpretations are valid and which is the deepest.

Here is what perhaps the greatest physics professor, Richard Feynman, had to say on this topic:

> One of the amazing characteristics of nature is the variety of interpretational schemes which are possible. It turns out that it is only possible because the laws are just so, special and delicate. If you modify the laws much you find that you can only write them in fewer ways. I always find that mysterious, and I do not understand the reason why it is that the correct laws of physics seem to be expressible in such a tremendous variety of ways. They seem to be able to get through several wickets at the same time.[1]

What's going on? Why does looking for unification and simplicity work so well? And why are the laws of nature so simple and unifying, yet expressible in many ways? Because that's how Creation was done:

> Creation, as described in the mystical teachings, is an evolution from the utterly singular to the plural and dichotomous. ... [Thus] the deeper you delve, descending to the molecular, atomic, and subatomic levels, the more unanimity you will find. [written circa 1900][2]

Why do scientists intuitively look for these unifying concepts? Because this drive is inherent to the souls of scientists—and all humanity. Humans consist of a body with similar functions to those of other animals. This body is activated by an animal soul; think of it as the software that runs the hardware, which is the body. The animal soul is the operating system, focused on survival. It keeps everything functioning and is responsible for our instincts. But in addition, humans have a divine soul,[3] like a series of apps with unique features, which science has discovered and categorized as modern human behaviors.[4] One of these features is that humans have a creative process similar to God's. As we saw earlier in this book, His creative process is to evolve things from the utterly singular. Hence, we instinctively or "religiously" look for the utterly singular behind all physical laws.

Unity in Nature[5]

Human knowledge is limited, and human intellect cannot visualize what is beyond it. Therefore, human reason cannot grasp the true essence of God because He is perfect, whereas human reason is imperfect. We can strive to understand divinity to the limit of our intellectual capacities; beyond that limit, we're left to believe with simple faith. The central idea, the pivot around which the concepts of creation revolve, is the principle of *Yichud HaShem*, Divine Unity.

The proclamation of this unity is made daily by Jews in the scriptural utterance "Shma Yisrael," "Hear, O Israel, G-d is our L-rd, G-d is One."[6] The meaning of this principle is understandable on many levels.

The essential meaning of the doctrine of Divine Unity is the belief in absolute monotheism—i.e., there is but one God, with none other besides Him. It negates polytheism, the worship of many gods, and paganism, the deification of any finite thing or being or natural force. It also excludes dualism, the assumption of two rival powers of good and evil, and pantheism, which equates God and nature.

Maimonides' interpretation of God's Unity also emphasizes that His essence and being is a simple and perfect unity without any plurality, composition, or divisibility, and that it is free from any physical properties and attributes. In his words, "none of the things existing in the universe to which the term one is applied is like unto His Unity; such as a physical body which consists of parts. There is no other unity like it in the world."[7]

The Chassidic interpretation[8] gives it a more profound meaning: Divine Unity not only excludes the existence of other ruling powers besides the one God, or of any plurality in Him, but it also precludes any existence at all apart from Him. The universe appears to possess an existence independent from its Creator only because we do not perceive the creating force, which is its *raison d' être*.[9] This interpretation of Divine Unity is closely related to the other fundamental doctrine we've encountered in this book—the doctrine of tzimtzum.

Jewish philosophic and Kabbalistic thought has been deeply concerned with the seeming contradiction between the divine attribute of omnipresence and the existence of the universe. Since God is omnipresent and nothing can exist outside of Him, where was there a place for the universe to materialize at Creation? How could the finite world emerge from God the infinite? How can things that are the antithesis of divinity exist in His presence?

As we have seen, this was explained by Rabbi Yitzchak Luria with the doctrine of tzimtzum: before the creation of the worlds, God filled all "space," and there was no room, no possibility, for the existence of the universe. However, when it arose in His will to create the world, the infinite withdrew to the sides, as it were, and a vacuum and empty space was formed, thereby making room for the universe to exist. Into this space there emanated from God a ray of light, the Kav, which is the source of all creation. This light and creating force underwent countless tzimtzumim, contracting and diminishing it until the corporeal and mundane world was brought into existence, containing even things that defy holiness.

Through this screening and concealment of the power by which all things exist, the world appears to enjoy an independent existence, as if it were apart from God. This gives us free will. But in truth, "The whole earth is full of His Glory."[10] Were permission granted to the eye to see beyond the external physical form, declares Rabbi Schneur Zalman, then we would perceive only the divine power that pervades and animates every created thing and is its true essence and reality.

Thus, behind nature there is complete unity, and science has been discovering and continues to discover that unity. How far will we go in understanding the unity via the scientific method? Only as far as the first out of nothing creation—only as far as the launching of the Big Bang, but not before it. In the absence of the eye being granted the power to see beyond external physical forms, we are limited to exploring *the first instant*, when the universe came out of nothing, to gain a glimpse of the Creator.

THANK YOU FOR READING THIS BOOK!

If you enjoyed *Mysteries of the First Instant* or found it useful, I'd be very grateful if you'd post a short review on Amazon. Your support really does make a difference! I read all the reviews personally so I can receive your feedback and make this book even better.

Thanks again for your support!

HAVE A QUESTION FOR DANIEL?

Did reading this book raise more questions in your mind?

Do you need clarification on any of the points outlined in this work?

Then head over to my website at danielfriedmannbooks.ca to ask your questions directly.

ABOUT THE AUTHORS

Daniel Friedmann, P.Eng., M.A.Sc., a Readers' Favorite International Book Award Winner, studies the origin of the universe and life on Earth. He is an expert on the creation–evolution debate. He is Chairman of Carbon Engineering, a company dedicated to removing CO_2 from the air to solve climate change. He was the President and CEO of a global communications and information company until May 2016 and is a student of religion. His work on reconciling the biblical account with scientific observation using his "biblical clock" formula has been reported in conferences, newspapers, and magazines, as well as on television and radio programs. He is the author of *The Genesis One Code*, *The Broken Gift*, *Roadmap to the End of Days*, and *The Biblical Clock*. Please visit danielfriedmannbooks.ca.

Dania Sheldon holds a doctorate in English language and literature from the University of Oxford and is a professional writer, editor, and researcher, working in a wide range of subject areas and genres across the humanities and sciences. In 2017, Dania received the Tom Fairley Award for Editorial Excellence, Canada's highest recognition in that profession. She is the author of *The Book of Lua: Stories and Wisdom from a Little Cat with Mobility Challenges*, short-listed for two Next Generation Indie Book Awards. Please visit www.daniasheldon.com and www.thebookoflua.com.

DOWNLOAD BOOKS 1, 2, 3, and 4

in
DANIEL FRIEDMANN'S
Inspired Studies Series

Daniel Friedmann's fourth book, co-authored with Dania Sheldon, *The Biblical Clock* is a narrative describing Friedmann's quest for answers that produced the prior three books below. It is a standalone volume and in relating the story of discovery covers the essential materials in the earlier trilogy.

Download here: https://www.amazon.com/Biblical-Clock-Universe-Humanity-Inspired-ebook/dp/B07P93D46V/

In Daniel's first book, *The Genesis One Code*, he demonstrates an alignment between the dates of key events pertaining to the development of the universe and the appearance of life on Earth as described in Chapters 1 and 2 of Genesis, and the dates derived from scientific theory and observation.

Download here: https://www.amazon.com/Genesis-One-Code-Harmonizing-observation-ebook/dp/B00RFB4KTY/

Daniel's second book, *The Broken Gift*, follows and extends the scope of *The Genesis One Code* to include the appearance and early history of humans.

Download here: https://www.amazon.com/Broken-Gift-Harmonizing-biblical-scientific-ebook/dp/B00RM92JS8/

His third book, *Roadmap to the End of Days*, looks at history from a supernatural point of view to discover the events and timing of the End of Days.

Download here: https://www.amazon.com/Roadmap-End-Days-Demystifying-Eschatology-ebook/dp/B01N6ZBH7V/

READ AN EXCERPT FROM THE NEXT BOOK IN THE ORIGINS SERIES

Mysteries of the Origin of Animals: *Illuminating What Science Hasn't Answered about the Inception and Development of Animal Life*

Discover insights into the big open questions about life's origins while gaining a better understanding of biblical texts on creation.

Chapter 1 from *Mysteries of the Origin of Animals* is available before the endnotes in this book.

To be notified of the book's publication follow Daniel Friedmann here https://www.amazon.com/Daniel-Friedmann/e/B005LKD1Q4/

Or please go to https://www.danielfriedmannbooks.ca/contact/

Appendix A

The Standard Model

Particle Names and Properties[1]

Quarks

Name	Symbol	Charge (e)	Mass (MeV/c^2)	Spin
up	u	$+\frac{2}{3}$	$2.16^{+0.5}_{-0.3}$	$\frac{1}{2}$
top	t	$+\frac{2}{3}$	$172{,}760 \pm 300$	$\frac{1}{2}$
strange	s	$-\frac{1}{3}$	93^{+11}_{-5}	$\frac{1}{2}$
down	d	$-\frac{1}{3}$	$4.67^{+0.5}_{-0.2}$	$\frac{1}{2}$
charm	c	$+\frac{2}{3}$	1270 ± 20	$\frac{1}{2}$
bottom	b	$-\frac{1}{3}$	$4{,}180^{+30}_{-20}$	$\frac{1}{2}$

Leptons

Name	Symbol	Charge (e)	Mass (MeV/c^2)
Electron	e^-	−1	0.511
Electron neutrino	ν_e	0	< 0.0000022
Muon	μ^-	−1	105.7
Muon neutrino	ν_μ	0	< 0.170
Tau	τ^-	−1	1,776.86 ± 0.12
Tau neutrino	ν_τ	0	< 15.5

Bosons

Name	Symbol	Charge (e)	Spin	Mass (GeV/c^2)	Interaction mediated
Z boson	Z	0	1	91.1876 ±0.0021	Weak interaction
W boson	W^-	−1	1	80.379 ±0.012	Weak interaction
Photon	γ	0	1	0	Electromagnetism
Higgs boson	H^0	0	0	125.10 ±0.14	Mass
Gluon	g	0	1	0	Strong interaction

Appendix B

The Light Metaphor

The light metaphor[1] is used by the mystics to describe the various emanations and manifestations of the Divinity. Although by no means a perfect metaphor, light is chosen because its properties that are perceived by us as nonphysical make it analogous to God. For example:

(i) The existence of light cannot be denied.

(ii) Light is not a corporeal thing (as perceived by us without making careful physics experiments to reveal its wave and particle properties in detail).

(iii) Light when it interacts with the faculty of sight causes the visible colors to pass from potentiality to actuality.

(iv) Someone who has never seen a luminous body in their life cannot conceive of colors or the agreeableness of light.

(v) Even someone who has seen luminous objects cannot tolerate gazing upon intense light for any length of time, and if they insist upon gazing beyond their eyes' capacity to endure, they cannot thereafter see things that would previously have been visible.

Light also has numerous qualities characteristic of the divine emanations. For example:

(i) Light is emitted from the luminary without ever becoming separated from it. Even when its source is concealed or removed, thus no longer emitting perceptible light, the previous rays do not remain entities separate from the luminary but are withdrawn with it.

(ii) Light spreads itself in a way that seems to us to be instantaneous (although we know it has a finite speed).

(iii) Light irradiates all physical objects and can penetrate, unhindered, all transparent objects.

(iv) Light does not mix and mingle with any other substance.

(v) Light per se never changes. The perception of more or less intense light, or of differently colored lights, is not due to any change in the light per se but is due to external factors (filters).

(vi) Light is essential to life in general.

(vii) Light is received and absorbed relative to the capacities of the recipient.

However, we must remember that all terms and concepts related to God must be stripped of all and any temporal, spatial, and corporeal connotations and must be understood in a strictly spiritual sense. Thus, the light analogy is inadequate. For example, the emittance of perceptible light from its source is automatic and intrinsically necessary: the luminary cannot withhold the light. This restrictive quality cannot be ascribed to the emanations of the omnipotent and in fact are part of the creation process.

The Language Applied to God

In the Kabbalistic writings, God is referred to as the Ein Sof or the Infinite, He that is without limit. In the analogy, the Ein Sof is the luminary and radiator of light. This luminary radiates the Ohr Ein Sof, its light and radiation. Light in Hebrew is "ohr."

It is important to realize that in the analogy, the Ein Sof is omnipresent and never changes. It is the Ohr Ein Sof that changes. For example, the creation of physical space was brought about by a contraction or "withdrawal" and concentration of the Ohr Ein Sof: the omnipresent, Infinite Light of the Ein Sof, the Ohr Ein Sof, was screened, dimmed, hidden, and concealed, and where it was dimmed, an almost "empty" place, a "void" appeared—the original primordial space from which the universe expanded.

The Light Metaphor Extended

One can extend the light metaphor to understand how the Torah is the blueprint for Creation and in particular how the Hebrew letters (which God spoke) manifest reality. In this metaphor, the Ohr Ein Sof is analogous to the light from a film projector and the Torah to the film. Thus, the Godly light shines through the text of the Torah to display or create an image on the screen. This image is reality as we exist in it, and is created continuously; should the projector turn off, everything would cease and go back to nothing (in the analogy, a dark screen, in reality, no space or time).

Thus, in this analogy, the creation process consists of withdrawing the Infinite Light so the screen is visible (versus being washed out by the intensity of the light), and in this almost complete darkness, shining a dimmer (filtered) light though the Torah film to create reality on the screen. Of course, the analogy is not exact, but it's sufficiently useful.

Appendix C

The Hebrew Letters and Their Numerical Values

100	ק	Kuf	10	י	Yod	1	א	Alef
200	ר	Resh	20	ד,כ	Kaf	2	ב	Bet
300	ש	Shin	30	ל	Lamed	3	ג	Gimel
400	ת	Tav	40	ם,מ	Mem	4	ד	Dalet
			50	ן,נ	Nun	5	ה	Hei
			60	ס	Samech	6	ו	Vav
			70	ע	Ayin	7	ז	Zayin
			80	ף,פ	Pe	8	ח	Chet
			90	ץ,צ	Tzadi	9	ט	Tet

Glossary

A priori: Made before or without examination; not supported by factual study.

Abraham: Originally called Abram, he is given the name Abraham at the Covenant described in Genesis 17. God calls Abraham to leave his land, family, and household in Mesopotamia in return for a new land, family, and inheritance in Canaan, the Promised Land. Abraham's story ends with the death and burial of his wife Sarah in the grave that he has purchased in Hebron, followed by the marriage of his heir, Isaac, to a wife from his own people. These two episodes signify (1) the right of his descendants to the land and (2) the exclusion of the land's previous inhabitants, the Canaanites, from Israel's patrimony. For Jews, Abraham is the first of the three Patriarchs.

Adam: The first created man. He looked nothing like us. Only after his sin was he diminished (in size and intellect), thereby becoming more like us. Normal humans are referred to as humankind. Adam is referred to as Adam or man.

Age of the universe: How long our universe has existed—13.8 billion years according to scientific observations and the Big Bang theory, and 5,781 years according to the biblical timeline.

Anisotropy (of the CMB): The anisotropy of the cosmic microwave background (CMB) consists of the small temperature fluctuations in the blackbody radiation left over from the Big Bang. The average temperature of this radiation is 2.725 Kelvin.

Arizal (Rabbi Itzhak Luria): Rabbi Isaac Luria, known as the Arizal or Ari (1534–1572 CE) and considered the father of contemporary Kabbalah. His teachings are referred to as Lurianic Kabbalah; these describe new, coherent doctrines on the origins of Creation and its cosmic rectification, while incorporating a recasting and fuller systemization of preceding Kabbalistic teaching.

Arrow of time: The concept positing the "one-way direction" of time toward the future.

Asman: Hebrew for "time."

Atom: The smallest constituent unit of ordinary matter that constitutes a chemical element. Every solid, liquid, gas, and plasma is composed of atoms. They are extremely small; typical sizes are around a ten-millionth of a millimeter. Every atom is composed of a nucleus and one or more electrons bound to the nucleus.

Atomic clock: A clock regulated by the vibrations of an atomic or molecular system. Atomic clocks, also called primary clocks, are currently the most exact in the world. Most clocks measure time by counting how many times something moves back and forth. Atomic clocks count how many times the electrons of atoms inside the clock change energy levels.

Atomic nucleus: The small, dense region at the center of an atom, consisting of protons and neutrons.

Atomic number: The number of protons in the nucleus of an atom, which determines the chemical properties of an element and its place in the periodic table.

Bara: See creation.

BCE: Before the Common Era.

Beginning, the: Herein used as the beginning of our universe, the point when something appeared from nothing.

Ben: Hebrew for "son of."

Bible: See Torah.

Biblical timeline: The three-phase timeline for the universe, comprised of the six days of Creation, followed by the 6,000 years of human history, and the seventh millennium. On the sixth day, Adam

Glossary

was made. This marked the start of the first year of Creation, and since then, 5,781 years have elapsed as of 2021. History will last till the biblical year 6000 or earlier, after which will come the seventh millennium.

Biblical years: Years as enumerated in the Bible, starting at year 1 after Day 6 of Creation and continuing to the year 6000. Biblical years correspond to Gregorian calendar years as follows:

Biblical years	CE years
1	3760 BCE
1001	2760 BCE
2001	1760 BCE
3001	760 BCE
4001	241 CE
5001	1241 CE
6000	2240 CE

Big Bang: The prevailing cosmological theory of the universe's origins and development. According to the Big Bang theory, the universe was originally in an extremely hot and dense state, then expanded rapidly. It has since cooled by continuing to expand to its present diluted state. Based on the best available measurements, scientists have determined the original state of the universe existed about 13.8 billion years ago. This theory offers the most accurate, logical explanation supported by current scientific evidence and observations.

Blueprint: A design plan or other technical drawing for something to be built. Herein used in the context of the design/plan for the creation and building of everything in the universe.

Book of Creation: See *Sefer Yetzirah*.

Boson: An elementary particle that follows Bose–Einstein statistics. Examples of bosons include fundamental particles such as photons, gluons, and W and Z bosons (the four force-carrying particles of the

Standard Model), the recently discovered Higgs boson, and the hypothetical graviton of quantum gravity.

Canadarm: Canadarm 1 was a remote-controlled mechanical arm, also known as the Shuttle Remote Manipulator System (SRMS). During its thirty-year career with NASA's Space Shuttle Program, the robotic arm deployed, captured, and repaired satellites, positioned astronauts, maintained equipment, and moved cargo. Canadarm 2 is the larger robotic arm permanently at the International Space Station. It is used for berthing the trusses and commercial vehicles and for inspecting the whole space station.

Cantonists: Used to refer to young Jewish men kidnapped from their parents by the czarist regime and conscripted for twenty-five years or more in the Russian army.

Chalal: The first physical creation and primordial space left after the first contraction, or tzimtzum, of the Ohr Ein Sof, which literally means vacated space.

Chassidic movement: A Jewish Orthodox spiritual revivalist movement that emerged in Eastern Europe in the eighteenth century. Followers of Hasidic Judaism (known as Hasidim, or "pious ones") drew heavily on the Jewish mystical tradition, seeking a direct experience of God through ecstatic prayer and other rituals conducted under the spiritual direction of a rebbe.

Classical mechanics: Describes the motion of macroscopic objects, from projectiles to parts of machinery, and astronomical objects, such as spacecraft, planets, stars, and galaxies.

Classical physics: Refers to theories of physics that predate modern, more complete, or more widely applicable theories. It includes classical mechanics.

CMB: See cosmic microwave background.

Commandments: 613 commandments given by God to the Jewish people (for a list, see http://www.jewfaq.org/613.htm).

Glossary

Commentaries: Critical explanations or interpretations of biblical texts.

Copenhagen interpretation: An expression of the meaning of quantum mechanics that was largely devised from 1925 to 1927 by Niels Bohr and Werner Heisenberg. It is one of the oldest of numerous proposed interpretations of quantum mechanics and remains one of the most commonly taught. According to the Copenhagen interpretation, physical systems generally do not have definite properties prior to being measured, and quantum mechanics can only predict the probability distribution of a given measurement's possible results. The act of measurement affects the system, causing the set of probabilities to reduce to only one of the possible values immediately after the measurement. This feature is known as wave function collapse. If there is no observation, this collapse does not occur, and none of the options ever becomes less likely.

Cosmic microwave background (CMB): Electromagnetic radiation remnants from an early stage of the universe. The CMB is faint radiation filling all space. It is an important source of data on the early universe because it is the oldest electromagnetic radiation. With a traditional optical telescope, the space between stars and galaxies (the background) is completely dark. However, a sufficiently sensitive radio telescope shows a faint background glow that is not associated with any star, galaxy, or other object—this is the CMB.

Cosmic time: The time coordinate commonly used in the Big Bang theory of physical cosmology. It is defined for a homogeneous expanding universe as follows: (1) choose a time coordinate in which the universe has the same density everywhere at each moment in time; (2) measure the passage of time using clocks that move with the expanding universe; and (3) place the Big Bang singularity (i.e., time zero) at the center.

Cosmological parameters: Parameters that define the properties of the universe and are the main input for the Big Bang theory.

Cosmological principle: The principle that, "viewed on a sufficiently large scale, the properties of the universe are the same for all observers." This means the spatial distribution of matter in the universe is homogeneous (the same in all locations) and isotropic (the same in all directions).

Cosmology: The study of how the universe began and developed.

Creation (bara): The divine act of creating something from nothing.

Creation day: Equivalent to 2.54 billion years in our time.

Curvature of space: The amount by which a curve deviates from being a straight line, or a surface deviates from being a plane, and by extension the curvature of three-dimensional space or four-dimensional space–time. According to general relativity, the curvature of space–time—which governs the motion of objects in the universe, including radiation (light)—is directly related to the energy and momentum of whatever matter and radiation are present.

Dark energy: A hypothetical energy and pressure, uniformly filling space, that causes a repulsive force and the universe's expansion to accelerate. It can vary with time.

Dark matter: Matter throughout space that exerts attractive gravity but does not emit light, and is not detectible by methods used to detect normal or visible matter.

Design manual: A document that tells you how to build something, herein used in the sense of the manual to create and build the universe.

Deuterium: Also known as heavy hydrogen, this is an isotope of hydrogen. The nucleus of a deuterium atom contains one proton and one neutron, whereas the far more common hydrogen has no neutron in the nucleus. Deuterium has a natural abundance in the Earth's oceans of about one atom in 6,420 atoms of hydrogen.

Divine Plan: God's plan for history and humanity.

Divine purpose: To make the physical world a dwelling place for God.

Glossary

Ecliptic, plane of: The imaginary plane containing the Earth's orbit around the sun.

Electroweak theory: A theory unifying the weak and electromagnetic forces.

Electromagnetic force: One of the four fundamental forces. It is an explanation for how both moving and stationary charged particles interact. It's called the electromagnetic force because it includes the formerly distinct electric force and magnetic force; magnetic forces and electric forces are really the same fundamental force. The electric force acts between all charged particles, whether or not they're moving. The magnetic force acts between moving charged particles.

Electron: A subatomic particle whose electric charge is negative one elementary charge. Electrons belong to the first generation of the lepton particle family. The electron has a mass that is approximately $1/1,836$ that of the proton.

Electron volts: A unit of energy (or, since energy and mass are equivalent, of mass) equal to the work done on an electron when accelerating it through a potential difference of one volt.

Elementary particle: A subatomic particle with no substructure, meaning it is not composed of other particles. Particles currently thought to be elementary include the fundamental fermions (quarks, leptons), which generally are matter particles, as well as the fundamental bosons (gauge bosons and the Higgs boson), which generally are force particles.

Element: An element is a substance whose atoms all have the same number of protons. Elements are chemically the simplest substances and hence cannot be broken down using chemical reactions. Elements are displayed tabularly in the periodic table.

Elokim: The name of God used in Chapter 1 of Genesis, indicating that the actions in the Creation account were governed by strict law and order, and everything that occurred had to be based on cause and effect.

Ein Sof: Refers to God in the analogy of light. It means the Infinite, He that is without limit.

End of Days: Also called End Time or End Times, end of time, last days, final days, or eschaton, it is a time period concerned with the final events in history, described in the theologies of the dominant world religions, both Abrahamic and non-Abrahamic.

Etz Hayim: *The Tree of Life*, a collection of the Arizal's teachings written by Rabbi Chaim Vital. In eight volumes, it became the core text synthesizing what is now known as Lurianic Kabbalah.

Fermions: Particles that include all quarks and leptons, as well as all composite particles made of an odd number of these, such as many atoms and nuclei. A fermion can be an elementary particle, such as the electron, or it can be a composite particle, such as the proton.

Field equations: Einstein equations; see general relativity.

Fine-tuning problem: The problem that arises from varying, by a very modest amount, almost any of the particular properties of the universe, the laws of physics, or their parameters, as the variation leads to circumstances whereby our universe cannot exist.

First instant: Herein used to describe the very short time period at the beginning of the universe (preceding the Big Bang) when something, including time and space, appeared out of nothing.

Flatness problem: A cosmological fine-tuning problem within the Big Bang theory of the universe. In this case, the fine-tuned parameter is the density of matter and energy in the universe. This value affects the curvature of space–time, with a very specific critical value being required for a flat universe, as it appears at the current time.

Forces of nature: The four fundamental forces that govern the universe: gravity, electromagnetic, weak nuclear force, and strong nuclear force.

Formation: The act of taking something that already exists and making it into something else.

Glossary

Friedmann equations: A set of equations in physical cosmology that govern the expansion of space in homogeneous and isotropic models of the universe, within the context of general relativity. They were first derived by Alexander Friedmann in 1922 from Einstein's field equations of gravitation.

Galaxy: A system of millions or billions of stars, together with gas and dust, held together by gravitational attraction. Galaxies can have various shapes, including flat spirals and oblate spheroids (ellipticals).

General relativity: The currently accepted theory of gravitation that was developed by Albert Einstein between 1907 and 1915. According to general relativity, the observed gravitational effect between masses results from their warping or curving of space–time.

Gluon: An elementary particle that acts as the exchange particle (force particle) for the strong nuclear force between quarks, and holds protons, neutrons, and atomic nuclei together. There are eight independent types of gluon.

God (essential name of): YHWH, called the tetragrammaton or essential name of God, or in Hebrew yod-hei-vav-hei. Other names of God refer to one of his manifestations in the world, such as discipline.

Grand unified theory (GUT): A model in particle physics in which, at high energy, the electromagnetic, weak, and strong forces are merged into a single force.

Gravity (force): A natural phenomenon by which all things with mass or energy—including planets, stars, galaxies, and even light—are brought toward (or gravitate toward) one another.

Hasidic Judaism: A movement within Judaism. It arose as a spiritual revival movement in contemporary Western Ukraine during the eighteenth century and spread rapidly throughout Eastern Europe. Today, it is worldwide. Hasidic thought draws heavily on Lurianic Kabbalah.

Hebrew: Herein, the language of the Torah, also known as the Holy Tongue. According to tradition, the twenty-two letters of the Hebrew alphabet hold the secrets to the Creation and the key to God's wisdom. Modern Hebrew is the official language of the state of Israel and differs in many respects from biblical Hebrew; for example, there are around 8,000 different Hebrew words in the Bible, while Modern Hebrew has over 100,000 words.

Higgs boson: An elementary particle in the Standard Model of particle physics. It is named after physicist Peter Higgs, who in 1964, along with five other scientists, proposed the Higgs mechanism to explain why particles have mass. This mechanism implies the existence of the Higgs boson. The boson's existence was confirmed in 2012 based on particle collisions in the Large Hadron Collider at CERN.

Homogeneous: The same in all locations

Horizon problem: A cosmological fine-tuning problem within the Big Bang theory of the universe. The horizon problem arises due to the difficulty in explaining the observed homogeneity of causally disconnected regions of space, in the absence of a mechanism that sets the same initial conditions everywhere.

Hubble key project: A project using the Hubble Space Telescope to determine the Hubble Constant by the systematic observations of Cepheid variable stars, which brighten and dim periodically, allowing them to be used as cosmic yardsticks up to distances of a few tens of millions of light years.

Hubble space telescope: A space telescope that was launched into low Earth orbit in 1990 and remains in operation. Although not the first space telescope, Hubble is one of the largest and most versatile and is well known as a vital research tool. It is named after the astronomer Edwin Hubble.

Inflation: Cosmic inflation, cosmological inflation, or just inflation is a theory positing the exponential expansion of space in the early universe. It proposes that the inflationary epoch lasted an infinitesimal period, from 10^{-36} seconds after the conjectured singularity to sometime between 10^{-33} and 10^{-32} seconds after the singularity. Following the inflationary period, the universe continued to expand normally per the hot Big Bang theory. The detailed particle physics mechanism responsible for inflation is unknown.

Interpretation of quantum mechanics: See measurement problem.

Inter-inclusion: The concept that each channel of divine energy possesses within its own internal makeup something of all the other channels and is thus complete.

Isaac ben Samuel of Acre (late 13th to mid-14th century): A Kabbalist who lived in the land of Israel; author of *Otzar HaChaim*.

Isotropic: The same in all directions.

It was good: Phrase meaning "completed to the point that it was useful to man."

Kabbalah: Hebrew for "receiving" or "tradition," it is a discipline and school of thought concerned with the mystical aspect of Judaism.

Kav: The ray of light (meaning the Ohr Ein Sof) that emanated from God into the original void, which was the source of all creation.

Lego: A line of plastic construction toys that are manufactured by The Lego Group. The company's flagship product, Lego, consists of colorful interlocking plastic bricks accompanying an array of gears, figurines called minifigures, and various other parts. In this book, when we refer to Lego we mean only the interlocking bricks. Lego pieces can be assembled and connected in many ways to construct objects, including buildings. Anything constructed can be taken apart again, and the pieces reused to make new things.

Lepton: An elementary particle that does not undergo strong interactions. The two main classes of leptons are charged (also known as the electron-like leptons) and neutral (better known as neutrinos). Charged leptons can combine with other particles to form various composite particles such as atoms, while neutrinos rarely interact with anything and are consequently rarely observed. The best known of all leptons is the electron.

Letters (Hebrew): The twenty-two letters of the Hebrew alphabet with which the universe was "spoken" into existence. Thus, they are the building blocks of the universe.

LHC at CERN (Large Hadron Collider at the European Organization for Nuclear Research): The world's largest and highest-energy particle collider, and the largest machine in the world. It lies in a tunnel 27 kilometers (17 mi) in circumference and as deep as 175 meters (574 ft). The aim of the LHC's detectors is to allow physicists to test the predictions of different theories of particle physics, including the discovery and properties of the Higgs boson.

Light metaphor: The use of light as a metaphor to describe the various emanations and manifestations of the divinity.

Light, speed of: The speed at which light and massless particles move, approximately 300,000 kilometers per second. It's the fastest possible speed within space, although space itself can and does expand faster than the speed of light. According to special relativity, it is the upper limit for the speed at which conventional matter and information can travel. In the special relativity equation, it interrelates the mass–energy equivalence: $E = mc^2$, where E is energy, m is mass, and c is the speed of light.

Life force: See spark, divine.

Lubavitcher Rebbe: Menachem Mendel Schneersohn (April 5, 1902– June 12, 1994), a prominent Hasidic rabbi and the seventh and last Rebbe (Hasidic leader) of the Chabad–Lubavitch. Chabad–Lubavitch is a branch of Orthodox Judaism that promotes spirituality and joy through the popularization and internalization of Jewish mysticism as the fundamental aspect of the Jewish faith movement.

Lurianic Kabbalah: See Arizal.

Macroscopic: Usually used to mean "visible to the naked eye." When applied to physical phenomena, the macroscopic scale describes things as a person can directly perceive them, without the aid of magnifying devices. This is in contrast to objects smaller than several tens of micrometers. Colloquially, although not quite correctly, used to describe the distinction between classical and quantum mechanics.

Measurement problem: In quantum mechanics, the unresolved issue of what constitutes the macroscopic world and what constitutes the quantum world. Quantum mechanics predicts the probability of something happening. But when we try to measure something, say, the position of a particle, we always get a value. What happens in the measurement process to convert the quantum world to the actual macroscopic definite observation is not understood. The inability to observe the measurement process directly has given rise to different interpretations of quantum mechanics and poses a key set of questions that each interpretation must answer.

Messianic Era: An age when, under the leadership of the Messiah, the whole world will believe in one God and live together in peace and brotherhood; expected to start on or before the year 6000, or 2240 CE.

Microscopic: Usually used to mean so small as to be visible only with a microscope. However, in physical phenomena, it is associated with quantum mechanics. See macroscopic.

Midrash: Hebrew for "exposition." Denotes non-legalistic teachings of the rabbis of the Talmudic era. The plural for midrash is midrashim.

Midrash Rabbah: Midrash dedicated to explaining the Five Books of Moses.

Milky Way: The galaxy containing our solar system and consisting of a central older bulge and a younger disk; the latter is where the solar system formed.

Molecule: An electrically neutral group of two or more atoms held together by chemical bonds.

Moses: The greatest prophet of all time. Born in Egypt and raised by Pharaoh's daughter, he fled to Midian, where he married Zipporah. He was sent by God back to Egypt to liberate the Israelites. Moses visited ten plagues upon Egypt, led the Israelites out, and transmitted to them the Torah at Mt. Sinai. Then he led the Israelites for forty years while they wandered in the desert. He died in the Plains of Moab after writing the Five Books of Moses, and was succeeded by his disciple Joshua, who led the Israelites into the Promised Land.

Multiverse: A hypothetical group of multiple universes. In 1952, Erwin Schrödinger gave a lecture in which he jocularly warned his audience that what he was about to say might "seem lunatic." He said that when his equations seemed to describe several different histories, these were "not alternatives, but all really happen simultaneously." The multiverse is used to try to do away with the fine-tuning problem. If billions of universes exist, then they may all have different values for the key variables that determine the universes' characteristics. Thus, we live in one of those universes, and ours randomly developed the exact parameters for our existence.

Names of God: Different names that refer to various ways in which He reveals Himself.

Neutron: A subatomic particle with no net electric charge and a mass slightly greater than that of a proton. It is made up of three quarks from the lowest energy level.

Nothing (in Kabbalah): Nothing physical—no space, time, particles, or forces.

Nothing (in science): Often something, at least space and gravity. Sometimes referred to as the quantum vacuum.

Nuclear force, strong: A force that acts between the protons and neutrons of atoms. Neutrons and protons are affected by the nuclear force almost identically. Since protons have charge, they experience an electric force that tends to push them apart, but at short range, the attractive nuclear force is strong enough to overcome the electromagnetic force.

Nuclear fusion: A reaction in which two or more atomic nuclei are combined to form one or more different atomic nuclei and subatomic particles (specifically, neutrons or protons). The difference in mass between the reactants and products is manifested as the release of energy. Nuclear fusion powers stars.

Glossary

Nucleosynthesis: The process by which heavier chemical elements are synthesized from hydrogen nuclei in the interiors of stars.

Ohr Ein Sof: The radiance of God in the analogy of light. In Hebrew, "ohr" means light.

Order of time: A spiritual time or sequential plan that establishes the full history of the universe, with its various phases, according to the biblical timeline. It is different from physical time, which we experience as measured motion, such as the rotation of the Earth indicating a day.

Oral Law or Oral Torah: Used to interpret and apply the Written Law. It is now documented in writing. It consists primarily of the Talmud, Explanations, Midrashim, and Zohar.

Oral tradition: See Oral Torah.

Organic compounds: Any chemical compounds that contain carbon. Due to carbon's ability to form chains with other carbon atoms, millions of organic compounds are known. They usually come from organisms, although sometimes they can be made synthetically.

Otzar HaChaim: Kabbalistic work by Isaac ben Samuel. It was the first work to state that the universe is actually billions of years old. Isaac ben Samuel arrived at this conclusion by distinguishing between earthly solar years and divine years.

Particle accelerator: An apparatus for accelerating subatomic particles to high velocities by means of electric or electromagnetic fields. The accelerated particles are generally made to collide with other particles as a research technique.

Periodic table: A table of the chemical elements arranged in order of atomic number, usually in rows, so that elements with similar atomic structure (and hence similar chemical properties) appear in vertical columns.

Photon: A type of elementary particle. It is the quantum of the electromagnetic field, including electromagnetic radiation such as light and radio waves, and the force carrier for the electromagnetic force. A photon has zero mass, and in a vacuum it always moves at the speed of light.

Positron: The antiparticle or antimatter counterpart of the electron. The positron has the same mass as an electron but the opposite charge.

Proton: A subatomic particle with a positive electric charge of one elementary charge and a mass slightly less than a neutron's. Protons and neutrons are collectively referred to as "nucleons." One or more protons are present in the nucleus of every atom. The number of protons in the nucleus is the defining property of an element and is referred to as the atomic number.

Quantized: Refers to a variable quantity restricted to discrete values rather than to a continuous set of values. For example, electron spin can be $+\frac{1}{2}$ or $-\frac{1}{2}$, nothing else.

Quantum mechanics: A fundamental theory in physics that describes nature at the smallest scales of atoms and subatomic particles.

Quantum vacuum: That very early state of the universe in the first fraction of a second when it was so hot and dense that physical particles could not exist. However, in the present-day understanding of the quantum vacuum, it was by no means a simple empty space. According to quantum mechanics, the vacuum state was not truly empty but instead contained fleeting electromagnetic waves and particles that popped into and out of existence. So in this state, time, space, the laws of physics, and particles existed, albeit at such a temperature that as soon as they appeared, they turned back into energy.

Quark: A type of elementary particle and a fundamental constituent of matter. Quarks combine to form composite particles called hadrons, the most stable of which are protons and neutrons, the components of atomic nuclei. Quarks are never directly observed or found in isolation. They have various intrinsic properties, including electric charge, color charge, mass, and spin.

Rabbi Isaac Luria: See Arizal.

Glossary

Radioactive dating: Measurement of the amount of radioactive material that an object contains; used to estimate the object's age.

Rambam: Rabbi Mosheh ben Maimon, called Moses Maimonides and also known by his acronym RaMBaM (1135–1204 CE), was a preeminent medieval philosopher and astronomer and one of the most prolific and influential Torah scholars and physicians of the Middle Ages in Morocco and Egypt.

Ramban: Nachmanides, also known as Rabbi Moses ben Nachman Girondi or as Bonastrucça Porta, and by his acronym Ramban (1194–1270 CE); a leading medieval scholar, rabbi, philosopher, physician, and biblical commentator.

Rashi: Shlomo Yitzhaki (1040–1105 CE), better known by the acronym Rashi (RAbbi SHlomo Itzhaki), was a medieval French rabbi famed as author of the first comprehensive commentary on the Talmud as well as a comprehensive commentary on the Written Law (including Genesis).

Rebbe: The spiritual leader in the Hasidic movement.

Relativistic speed: A speed at which effects predicted by special relativity manifest. It is close to the speed of light.

Relativity: Two interrelated theories developed by Albert Einstein—special relativity and general relativity. Special relativity applies to all physical phenomena in the absence of gravity and when things move at near the speed of light. General relativity explains the law of gravitation and applies to the cosmological and astrophysical realms. The theory of relativity transformed theoretical physics during the twentieth century, superseding a 200-year-old theory of mechanics created primarily by Isaac Newton. With relativity, cosmology and astrophysics predicted extraordinary astronomical phenomena such as neutron stars, black holes, and gravitational waves.

Sabbatical year: In Hebrew this is *shmita*, which literally means "release." It is the seventh year of the seven-year agricultural cycle mandated by the Torah for the Land of Israel.

Schrödinger: Erwin Rudolf Josef Alexander Schrödinger (August 12, 1887 – January 4, 1961) was a Nobel Prize-winning Austrian physicist who developed a number of fundamental results in the field of quantum theory.

Science: The systematic process of gathering information about the world and organizing it into theories and laws that can be tested.

Scientific method: A system of processes used to establish new or revised knowledge.

Seder asman: See order of time.

Sefer Yetzirah: The oldest Kabbalistic text. The book is attributed to Abraham. It was then passed on orally until it was redacted by Rabbi Akiva.

Sefirah (plural sefirot): A channel of divine energy or life force. There are ten sefirot. It is via the ten sefirot that God interacts with creation; they may thus be considered His attributes.

Sefirot: See sefirah.

Seventh millennium: The beginning of the era of universal reward known as the World to Come, which comes after the End of Days, or the Messianic Era. The seventh millennium is also referred to as The Millennium by some Christian denominations.

Shabbat: A day of rest for Jews, specifically the seventh day of the week, which begins on Friday at sunset and ends on the following evening after nightfall. It is observed in honor of the biblical Creation. On Shabbat, Jews exercise their freedom from the regular labors of everyday life; it offers an opportunity to contemplate the spiritual aspects of life and to spend time with family.

Singularity: A point of infinitely small space and infinite density where all the laws of physics break down.

Space Shuttle: A partially reusable, low Earth orbital spacecraft system operated from 1981 to 2011 by NASA as part of the Space

Glossary

Shuttle Program. Operational space shuttle missions launched, repaired, and upgraded the Hubble Space Telescope.

Spark, divine: Small spark of divine light that provides the life force of anything that exists—from what keeps an elementary particle in existence, to the human soul.

Special relativity, theory of: See relativity.

Standard Model of particle physics: The theory describing three of the four known fundamental forces in the universe and classifying all known elementary particles. It was developed in stages throughout the latter half of the twentieth century. Although the Standard Model is believed to be theoretically self-consistent and has demonstrated huge successes in providing experimental predictions, it leaves some phenomena unexplained and falls short of being a complete theory of fundamental interactions. For example, it does not: incorporate the full theory of gravitation as described by general relativity; account for the accelerating expansion of the universe as possibly described by dark energy; or contain any viable dark matter particle that possesses all of the required properties deduced from observational cosmology.

Star: An astronomical object consisting of a luminous spheroid of plasma held together by its own gravity. The nearest star to Earth is the sun. The observable universe contains an estimated 1×10^{24} stars, but most are invisible to the naked eye from Earth, including all stars outside our galaxy, the Milky Way.

Structure problem: A cosmological fine-tuning problem within the Big Bang theory of the universe. The structure problem arises because the Big Bang predicts a perfectly homogeneous universe in which galactic structures would not form. The fact that galaxies have been shown to cluster locally, with great voids between them, is proof of the inhomogeneity of the universe. Moreover, anisotropies have been found in the cosmic microwave background. However, there is no mechanism within the Big Bang theory to account for these "seeds," or perturbations, that result in the large-scale structure.

Talmud: Hebrew for "instruction, learning," the Talmud is a central text of mainstream Judaism, in the form of a record of rabbinic discussions pertaining to Jewish law, ethics, philosophy, customs, and history.

Teli: A mysterious word found in the *Sefer Yetzirah*. It's believed to be the axis of the plane of the ecliptic. The teli is the center around which everything in the universe happens or is arranged—and even perhaps rotates.

This World: The 6,000 years of history post Eden, culminating in the End of Days. It is followed by the World to Come and preceded by the six days of Creation and existence in the Garden of Eden.

Tikkun: Hebrew for "rectification."

Tikkun, world of: Our macroscopic world, which is one of rectification. The world where classical physics applies.

Tohu: Hebrew for "chaos and formless."

Tohu, world of: The world that existed at the very first instant: a very thin, chaotic substance that everything was made from. Today, tohu refers to the microscopic world, where quantum mechanics applies.

Torah: Consists of both the Written Law and the Oral Law (see separate entry). The Written Law itself consists of the following:

    The Five Books of Moses (The Law):
- Bereishith (In the beginning) (Genesis)
- Shemoth (The names) (Exodus)
- Vayiqra (And He called) (Leviticus)
- Bamidbar (In the wilderness) (Numbers)
- Devarim (The words) (Deuteronomy)

    NEVI'IM (The Prophets):
- Yehoshua (Joshua)
- Shoftim (Judges)
- Shmuel (I &II Samuel)
- Melakhim (I & II Kings)
- Yeshayah (Isaiah)
- Yirmyah (Jeremiah)

- Yechezqel (Ezekiel)
- The Twelve (treated as one book)
  - Hoshea (Hosea)
  - Yoel (Joel)
  - Amos
  - Ovadyah (Obadiah)
  - Yonah (Jonah)
  - Mikhah (Micah)
  - Nachum
  - Chavaqquq (Habbakkuk)
  - Tzefanyah (Zephaniah)
  - Chaggai
  - Zekharyah (Zechariah)
  - Malakhi

KETHUVIM (The Writings):
- Divrei Ha–Yamim (The words of the days) (Chronicles)
- Tehillim (Psalms)
- Iyov (Job)
- Mishlei (Proverbs)
- Ruth
- Shir Ha–Shirim (Song of Songs)
- Qoheleth (the author's name) (Ecclesiastes)
- Eikhah (Lamentations)
- Esther
- Daniel
- Ezra and Nechemyah (Nehemiah) (treated as one book)

Twelve Tribes: The tribes formed mainly by the natural increase of the offspring of Jacob. The descendants of each of his sons are believed to have held together and thus constituted a social entity ("tribe").

Tzimtzum: A term used in the Lurianic Kabbalah to explain the doctrine that God carried out the process of Creation by "contracting" his Ohr Ein Sof (Infinite Light) to allow for a conceptual space or void in which finite and seemingly independent realms could exist via many more contractions, or tzimtzumim.

Tzemach Tzedek: Menachem Mendel Schneersohn (September 9, 1789 – March 17, 1866), the third Rebbe (spiritual leader) of the

Chabad–Lubavitch Hasidic movement. He assumed the leadership of Lubavitch on May 5, 1831. He was known as the Tzemach Tzedek ("Righteous Sprout" or "Righteous Scion"), after the title of a voluminous compendium of halakha (Jewish law) that he authored. He also authored *Derech Mitzvosecha* (Way of Your Commandments), a mystical exposition that includes many concepts relating to the first instant, particularly the nature of time.

Universe: All of space, time, and their contents, including planets, stars, galaxies, and all other forms of matter and energy. The spatial size of the entire universe is unknown.

Universe, observable: A spherical region of the universe comprising all matter that can be observed from Earth at the present time, because electromagnetic radiation from these objects has had time to reach the Earth since the beginning of the cosmological expansion. The word "observable" in this sense does not refer to the capability of modern technology to detect light or other information from an object, or whether there is anything to be detected; it refers to the physical limit created by the speed of light itself. It is possible to calculate the size of the observable universe, which is currently estimated to be ninety-three billion light years in diameter.

Universe, static: Also referred to as a "stationary" or "infinite" universe, this is a cosmological model in which the universe is both spatially infinite and temporally infinite, and space is neither expanding nor contracting. Albert Einstein proposed a variation of this universe: one that is temporally infinite but spatially finite. After the discovery that the universe is expanding, most scientists, including Einstein, accepted the Big Bang theory of an expanding universe with a beginning.

User's manual: A document that tells you how to use or operate something. Herein used to refer to the aspect of the Bible that tells us how to live morally.

Vessels: A Kabbalistic concept that God's light, the Ohr Ein Sof, must descend into a vessel, where it is contained. The original vessels from

Glossary

the world of tohu broke, so the divine light was then put into the vessels of tikkun, giving rise to the macroscopic world.

Vessels, broken: The vessels of tohu that broke; see vessels.

Visible matter: That which constitutes all directly visible mass (objects that emit or reflect electromagnetic radiation, whether light or other wavelengths, such as infrared). Visible matter includes stars, planet-sized objects, and smaller objects, made out of the elements in the periodic table.

Vital, Chaim: (October 11, 1542 – April 23, 1620) was the foremost disciple of Isaac Luria. He recorded much of his master's teachings.

W and Z bosons: The elementary particles that mediate the weak nuclear force. The Z boson is electrically neutral. One W boson has a positive charge, the other a negative charge. All three of these particles are very short-lived (i.e., a tiny fraction of a second).

Weak nuclear force (or weak interaction): The mechanism of interaction between subatomic particles that is responsible for the radioactive decay of atoms. Radioactive decay (also known as nuclear decay, radioactivity, radioactive disintegration, or nuclear disintegration) is the process by which an unstable atomic nucleus decays to a lighter nucleus by emitting radiation. A material containing unstable nuclei is considered radioactive.

World to Come: The time that comes after the End of Days, the seventh millennium being its first stage. It is preceded by This World.

Written Torah: See Torah.

Young Earth Creationism: A form of creationism that asserts the universe and life were created by direct acts of God during a relatively short period, sometime between 5,700 and 10,000 years ago.

Zohar: Hebrew for "splendor, radiance," it is the foundational work in the literature of Jewish mystical thought known as Kabbalah.

Chapter Resources

Chapter 1

Bellis, Mary. (March 17, 2017). "Ole Kirk Christiansen and the History of LEGO." ThoughtCo.com. www.thoughtco.com/ole-kirk-christiansen-lego-1991644

Cendrowicz, Leo. (January 28, 2008). "LEGO Celebrates 50 Years of Building." *TIME Magazine*. content.time.com/time/world/article/0,8599,1707379,00.html

Lego Group. (n.d.). "Godtfred Kirk Christiansen." LEGO® History. www.lego.com/en-us/themes/lego-history/articles/godtfred-kirk-christiansen-49e0f6b4cb6d4d7a8e0d93cf76746152

Lego Group. (n.d.). "The LEGO® System in Play." LEGO® History. www.lego.com/en-us/themes/lego-history/articles/lego-system-in-play-60d5efbce6cf46a78794f108b3c19bda

Lego Group. (August 10, 2012). *The LEGO® Story*. LEGOClubTV www.youtube.com/watch?v=NdDU_BBJW9Y

"Ole Kirk Christiansen." (December 2004). MIT Inventor of the Week archive. web.archive.org/web/20130527123958/http://web.mit.edu/invent/iow/christiansen.html

Wainwright, Oliver. (September 29, 2017). "Everything is Awesome! The Brick-tastic Brilliance of the new Lego House." *The Guardian*. www.theguardian.com/artanddesign/2017/sep/29/everything-is-awesome-lego-house-architecture-billund-denmark

Chapter 2

Belenkiy, Ari. (2013). "'The Waters I am Entering No One yet Has Crossed': Alexander Friedman and the Origins of Modern Cosmology." *Origins of the Expanding Universe: 1912–1932.* ASP Conference Series, Vol. 471. Edited by Michael J. Way and Deidre Hunter.

Harrison, E. (2003). *Cosmology—The Science of the Universe*, 2nd ed. Cambridge, UK: Cambridge University Press.

O'Connor, J. J., & Robertson, E. F. (1997). "Aleksandr Aleksandrovich Friedmann." School of Mathematics and Statistics, University of St. Andrews. www-history.mcs.st-andrews.ac.uk/Biographies/Friedmann.html

Planck Collaboration. (2015). "Planck 2015 Results. XIII. Cosmological Parameters." *Astronomy & Astrophysics* 594: A13. doi:10.1051/0004-6361/201525830

Rogers, Judd D., et al. (February 28, 2018). "An Absorption Profile Centred at 78 Megahertz in the Sky-Averaged Spectrum." *Nature.* arxiv.org/abs/1810.05912

Ryden, B. (2017). *Introduction to Cosmology*, 2nd ed. Cambridge, UK: Cambridge University Press.

Tropp, Eduard A., Frenkel, Viktor Ya., & Chernin, Artur D. (2006). *Alexander A. Friedmann: The Man Who Made the Universe Expand.* Cambridge, UK: Cambridge University Press.

Chapter 3

Lightman, Alan. (2006). *A Sense of the Mysterious: Science and the Human Spirit.* New York: Vintage Books.

NASA. (n.d.). Dark Matter. *Imagine the Universe!* Goddard Space Flight Center. imagine.gsfc.nasa.gov/science/objects/dark_matter1.html

Overbye, Dennis. (December 27, 2016). "Vera Rubin, 88, Dies; Opened Doors in Astronomy, and for Women." *New York Times*. www.nytimes.com/2016/12/27/science/vera-rubin-astronomist-who-made-the-case-for-dark-matter-dies-at-88.html

Rubin, Vera. (1995). "A Century of Galaxy Spectroscopy." *Astrophysical Journal, 451*: 419–428. adsabs.harvard.edu/abs/1995ApJ...451..419R

Scoles, Sarah. (October 4, 2016). "How Vera Rubin Confirmed Dark Matter." Astronomy.com. www.astronomy.com/news/2016/10/vera-rubin

Chapter 4

Dicke, Robert. (June 18, 1985). Transcript of Interview with Martin Harwit. Oral History Interviews, American Institute of Physics. www.aip.org/history-programs/niels-bohr-library/oral-histories/4572

Dicke, R. H., Peebles, P. J. E., Roll, P. G., & Wilkinson, D. T. (July 1965). "Cosmic Black-Body Radiation." *Astrophysical Journal Letters*, 142(1): 414–419. doi:10.1086/148306. articles.adsabs.harvard.edu/pdf/1965ApJ...142..414D

"Fine-Tuning." (August 22, 2017). Stanford Encyclopedia of Philosophy. plato.stanford.edu/entries/fine-tuning/

Natarajan, Vasant, & Nityananda, Rajaram. (April 2011). "Robert H Dicke – Physicist Extraordinaire." *Resonance*, 16(4): 299–301. www.ias.ac.in/article/fulltext/reso/016/04/0299-0301

Nokia Bell Labs. (n.d.). "Cosmic Microwave Background Radiation." www.bell-labs.com/about/history-bell-labs/stories-changed-world/Cosmic-Microwave-Background-Discovery/

Peebles, P. J. E. (March 17, 1966). "Primeval Helium Abundance and the Primeval Fireball." *Physical Review Letters*, 16(2): 410–413. journals.aps.org/prl/pdf/10.1103/PhysRevLett.16.410

Penzias, A. A., & Wilson, R. W. (October 1965). "A Measurement of Excess Antenna Temperature at 4080 Mc/s." *Astrophysical Journal Letters*, 142(1): 419–421. http://articles.adsabs.harvard.edu/pdf/1965ApJ...142.1149P

Princeton University. (March 4, 1997). "Princeton Physicist Robert Dicke Dies." Communications and Publications, Stanhope Hall. https://pr.princeton.edu/news/97/q1/0304dick.html

Steinhardt, Paul J. (2011). "The Inflation Debate—Is the Theory at the Heart of Modern Cosmology Deeply Flawed?" *Scientific American* (April): 38–43.

Wall, Mike. (May 20, 2014). "Cosmic Anniversary: 'Big Bang Echo' Discovered 50 Years Ago Today." Space.com. https://www.space.com/25945-cosmic-microwave-background-discovery-50th-anniversary.html

Whittle, Mark. (2008). *Cosmology: The History and Nature of Our Universe. Course Guidebook.* Chantilly, VA: The Teaching Company.

Chapter 5

"Alan Guth." (2019). The Physics of the Universe. https://www.physicsoftheuniverse.com/scientists_guth.html

Greene, Brian. (2005). *The Fabric of the Cosmos: Space, Time, and the Texture of Reality.* New York: Vintage Books.

Guth, Alan. (1997). *The Inflationary Universe: The Quest for a New Theory of Cosmic Origins.* New York: Perseus Books.

Guth, Alan. (November 27, 2015). "Stumbling to Inflation" [Audio file]. The Story Collider: True Personal Stories about Science. https://www.storycollider.org/stories/2015/12/13/alan-guth-stumbling-to-inflation

Howgego, Joshua. (September 19, 2018). "10 Mysteries of the Universe: How Did It All Begin?" *New Scientist.* https://www.newscientist.com/article/mg23931960-200-10-mysteries-of-the-universe-how-did-it-all-begin/.

Chapter 6

CERN. (2014). *CERN and the Rise of the Standard Model* [video]. https://home.cern/science/physics/standard-model

National Institute of Standards and Technology. (n.d.). "Unit of Time (Second)." https://physics.nist.gov/cuu/Units/second.html

National Institute of Standards and Technology. (n.d.). "Unit of Length (Meter)." https://physics.nist.gov/cuu/Units/meter.html

"Peter Higgs." (n.d.). Notable Names Database. https://www.nndb.com/people/305/000169795/

Rodgers, Peter. (September 1, 2004). "The Heart of the Matter." *Independent*. https://www.independent.co.uk/news/science/the-heart-of-the-matter-54071.html

Rumbelow, Helen. (July 5, 2012). "Peter Higgs's Life Work a Heavy Burden to Bear." *The Australian*. https://www.theaustralian.com.au/news/world/higgs-a-heavy-burden-to-bear/news-story/c9bde99a9a6a6e331f8e6d70090304ae

Sample, Ian. (November 17, 2007). "The God of Small Things." *The Guardian*. https://www.theguardian.com/science/2007/nov/17/sciencenews.particlephysics

University of Edinburgh. (March 12, 2017). "Peter Higgs: Curriculum Vitae." School of Physics and Astronomy. https://www.ph.ed.ac.uk/higgs/peter-higgs

Chapter 7

Fine, Lawrence. (2003). *Physician of the Soul, Healer of the Cosmos: Isaac Luria and His Kabbalistic Fellowship*. Stanford, CA: Stanford University Press.

Friedmann, Daniel. (2017). *Roadmap to the End of Days*. Vancouver: Inspired Books.

Library of Congress. (2012). Words Like Sapphires: 100 Years of Hebraica at the Library of Congress, 1912–2012. https://www.loc.gov/exhibits/words-like-sapphires/cornerstones-of-jewish-religious-life.html

Mindel, Nissan. (n.d.). "Rabbi Isaac Luria – The Ari Hakodosh." Chabad.org. www.chabad.org/library/article_cdo/aid/111878/jewish/Rabbi-Isaac-Luria-The-Ari-Hakodosh.htm

Sherpin, Yehudah. (n.d.). "5 Works that Shaped Kabbalah." Chabad.org. https://www.chabad.org/library/article_cdo/aid/4018271/jewish/5-Works-That-Shaped-Kabbalah.htm

Chapter 8

de La Fuente, Marcos C., & de La Fuente, Marcos R. (1999). "Runaway Planets." *New Astronomy*, 4(1): 21–32.

Kaplan, Aryeh. (2008). *The Age of the Universe: A Torah True Perspective*. New York: Rueven Meir Caplan.

Mindel, Nissan. (n.d.). "Rabbi Shlomo Yitzchaki – Rashi." (n.d.). Chabad.org. www.chabad.org/library/article_cdo/aid/111831/jewish/Rabbi-Shlomo-Yitzchaki-Rashi.htm

NASA. (n.d.). "About the Hubble Space Telescope." www.nasa.gov/mission_pages/hubble/story/index.html

NASA. (n.d.). "How Old is the Sun?" Space Place. https://spaceplace.nasa.gov/sun-age/en/

"Rabbi Shlomo Yitzchaki – Rashi." (n.d.). Jewish Virtual Library. www.jewishvirtuallibrary.org/rabbi-shlomo-yitzchaki-rashi

Rashi. *Bereishit – Genesis – Chapter 1*. [English]. www.chabad.org/library/bible_cdo/aid/8165/jewish/Chapter-1.htm#showrashi=true

Solomin, Rachel M. (n.d.). "Counting the Jewish Years." My Jewish Learning. https://www.myjewishlearning.com/article/counting-the-years/

Chapter 9

Domnitch, Larry. (February 18, 2006). "The Cantonist Saga." Aish.com. https://www.aish.com/jl/h/h/48929772.html

"Rabbi Menachem Mendel, the 'Tzemach Tzedek' (1789–1866): A Brief Biography of the Third Chabad Rebbe." (n.d.). Chabad.org. https://www.chabad.org/library/article_cdo/aid/444/jewish/The-Tzemach-Tzedek.htm

Schneerson, Menachem Mendel. (n.d.). "Sand and Water." Translated and adapted by Yanke Tauber. https://www.chabad.org/library/article_cdo/aid/166403/jewish/Sand-and-Water.htm

Schneerson, Menachem Mendel. (April 4, 1985). "The Cantonists." [Video]. Chabad.org. https://www.chabad.org/therebbe/livingtorah/player_cdo/aid/1161502/jewish/The-Cantonists.htm

Chapter 10

None

Chapter 11

Ginsburgh, Yitzchak. (1990). *The Hebrew Letters: Channels of Creative Consciousness*. Jerusalem: Gal Einai Publications.

Friedmann, Daniel, & Sheldon, Dania. (2019). *The Biblical Clock: The Untold Secrets Linking the Universe and Humanity with God's Plan*. Vancouver: Inspired Books.

Kaplan, Aryeh. (1983). *The Aryeh Kaplan Reader. The Gift He Left Behind: Collected Essays on Jewish Themes from the Noted Writer and Thinker*. New York: Mesorah Publications.

Kaplan, Aryeh. (1997). "Introduction." *Sefer Yetzirah—The Book of Creation: In Theory and Practice*. San Francisco, CA: Weiser Books.

Langermann, Tzvi. (2017). "'Sefir Yesira,' the Story of a Text in Search of a Commentary." https://www.tabletmag.com/jewish-arts-and-culture/243868/sefer-yesira-text-commentary

Chapter 12

Ginsburgh, Yitzchak. (1990). *The Hebrew Letters: Channels of Creative Consciousness*. Jerusalem: Gal Einai Publications.

Kaplan, Aryeh. (1997). "Introduction." *Sefer Yetzirah—The Book of Creation: In Theory and Practice*. San Francisco, CA: Weiser Books.

Raskin, Aaron L. (n.d.). "Tav. The Twenty-Second Letter of the Hebrew Alphabet." Chabad.org. www.chabad.org/library/article_cdo/aid/137287/jewish/Tav.htm

IMAGE CREDITS

PART 1 image
Public domain
Image credit: NASA/WMAP science team
https://scienceblogs.com/startswithabang/2014/08/09/ask-ethan-49-do-the-cosmic-unknowns-cast-doubt-on-the-big-bang-synopsis

PART 2 image
Image credit: Red Milenaria
https://redmilenaria.com/etiquetas/cabala?page=12%2C7

PART 1

Figure 1.1
Public domain
Image credit: www.remodelaholic.com/free-vintage-printable-blueprints-diagrams/?m
G. K. Christensen, Billund, Denmark. Lego - 24. Oct. 1961.

Figure 1.2
Adapted from public data
Image credit: Science ABC
https://www.scienceabc.com/nature/universe/what-is-the-smallest-particle-we-know.html

Figure 2.1
Image credit: Wikimedia user "Decltype"
https://en.wikipedia.org/wiki/Einstein%27s_Blackboard#/media/File:Einstein_blackboard.jpg

Figure 2.2
Author created from public information

Figure 2.3
Author created from public information

Image Credits

Background image credit: Ethan Siegel
https://medium.com/starts-with-a-bang/ask-ethan-72-the-timeline-of-the-universe-9870d1b8f52

Figure 3.1
Public
Image credit: Pearson Education Inc. from http://beyondearthlyskies.blogspot.com/2013/05/black-holes-on-outskirts.html

Figure 3.2
Image credit: AIP Emilio Segrè Visual Archives, Rubin Collection; reproduced with permission.
https://photos.aip.org/history-programs/niels-bohr-library/photos/rubin-vera-f4

Figure 3.3
Image credit: Volker Springel, Virgo Consortium.
https://ucrtoday.ucr.edu/25894/cosmic-web

Figure 3.4
Author created from public information

Figure 5.1
Reproduced from
http://cds.cern.ch/record/823647/files/0502328.pdf?version=1

Figure 6.1
Image credit: CERN
https://home.cern/resources/image/accelerators/lhc-images-gallery

Figure 6.2
Adapted from public information
lhttps://www.physik.uzh.ch/groups/serra/StandardModel.html
Figure 6.3
Author created from public information

Figure 7.1
Public domain
michaeld.ca/zohar/1_zohar_mantua_1558.html

Figure 7.2
Image credit: Olive Seedlings Blog, January 2013,
shesileizeisim.blogspot.ca/2013/01/the-prohibition-of-onaas-devarim.html
3.bp.blogspot.com/-aQbExlSmtLM/UBK3ErTuYhI/
AAAAAAAAAS4/lIC7yL_4v44/s1600/ben+ish+chai.bmp

Figure 7.3
Public domain

Figure 7.4
Public domain

Figure 8.1
Public domain
Image credit: NASA
upload.wikimedia.org/wikipedia/commons/8/86/Andrew_Feustel_performs_work_on_the_Hubble_Space_Telescope.jpg

Figure 8.2
Author generated from public information

Figure 8.3
Public domain
Image credit: Jewish History Lectures
jewishhistorylectures.org/2013/11/19/his-155-1-7-the-talmud/

Figure 8.4
Public domain
www.geni.com/people/RASHI-%D7%A8%D7%A9-%D7%99/6000000067095013783

Figure 8.5
Author generated

Image Credits

Figure 9.1
Manuscript, *Etz Hayim* by the Ari, with the addition of various compilations.
Image credit: Kedem Auctions
www.google.com/search?q=24.%09Etz+Hayim+manuscript+circa+1770+image&safe=active&client=firefox-b&tbm=isch&source=iu&ictx=1&fir=jf-1aDtJuDi65M%253A%252CBBG7wPumOVx8KM%252C_&usg=AI4_-kTDiBLqFIIcyMgNIaxnaWFTCMaqbw&sa=X&ved=2ahUKEwisrtGQ9IbfAhWxOH0KHWPvCw4Q9QEwA3oECAUQCg#imgrc=jf-1aDtJuDi65M

Figures 9.2,3,4
Public domain
Credit: Chabad.org

Figure 9.5
Author generated

Figure 9.6
Author generated

Figure 10.1
Public domain
Image credit: Kennedy Warnken
www.bulbapp.com/u/cosmological-redshift

Figure 10.2
CC BY-SA 3.0
Image credit: Wikipedia
en.wikipedia.org/wiki/Ecliptic#/media/File:Earths_orbit_and_ecliptic.PNG

Figure 10.3
Public domain
Image credit: ESA and the Planck Collaboration
www.space.com/25945-cosmic-microwave-background-discovery-50th-anniversary.html

Figure 11.1
Public domain
Image credit: Schocken Institute
www.facebook.com/permalink.php?story_fbid=1166164193434038&id=199093986807735&pnref=story

Figure 11.2
Author generated

Figure 11.3
Author generated based on Harav Yitzchak Ginsburgh, *Lectures on Torah and Modern Physics* (Gal Einai, Israel, 2013), lecture 2

Figure 11.4
Public domain
Image credit: sciencesprings
sciencesprings.wordpress.com/tag/particle-physics/page/9/?iframe=true&preview=true%2Ffeed%2F

Figure 12.1
Public domain
Author-modified figure from *Sefer Yetzirah*

Figure 12.2
Created from public information
Background image
slideplayer.com/slide/3519958/

Figure 12.3
Public domain
Image credit: Janus Cosmological Model
januscosmologicalmodel.com/cptsymmetry

Figure 13.1
Author generated

Image Credits

PART 2

Figure 15.1
Public domain
Image credit: Awaken light
http://awakenlight.org/time-as-a-spiral

Figure 17.1
Author generated
Daniel E. Friedmann, "A Complete Set of 22 Elementary Particles for an Expanded Standard Model (Version 2)," *Open Access Library Journal*, 7 (September 2020): 1–10. 10.4236/oalib.1106715

EXCERPT
Mysteries of the Origin of Animals
Chapter 1

Trans-Canada Highway, just east of Field, British Columbia, July 2019

"In a whimsical convergence of science and Hollywood, a newly discovered crab species that lived 500 million years ago has been named after the legendary Millennium Falcon spaceship in *Star Wars*."

Reaching for the radio in my Jeep, I increased the volume and nudged my nephew, Seb, in the passenger seat. He remained slumped, but his droopy eyes opened wider in interest.

"Paleontologists in Canada have unearthed the fossil remains of *Cambroraster falcatus*, a crab with 'rake-like claws and a pineapple-slice-shaped mouth at the front of an enormous head,' which reached up to a foot in length. The fossils were found in the famed Burgess Shale area of the Rocky Mountains. Most animals alive at this time, which is known as the Cambrian Period, were 'smaller than your little finger,' says the study's lead author, Joe Moysiuk. So the discovery of this comparatively giant predator is an exciting development. 'This is not a sort of primitive, simple organism,' says Moysiuk. 'This is a highly specialized predator.' The first part of the crab's name refers to the Cambrian Period and the claws, while the second part comes from the striking resemblance its shell bears to the iconic spacecraft flown by the swashbuckling Han Solo.

"When asked about the naming, Harrison Ford—who played Han Solo in the movies and recently turned seventy-seven—is said to have laughed, stroked his silver-and-white beard, and responded, 'I think it's great. I'm glad they're naming a fossil after the spaceship and not after me. I feel old enough already!'"

The news moved on to another story, so I turned the volume down and said to Seb, "That's a neat coincidence, given that we're going to be at the Burgess Shale in a couple of days."

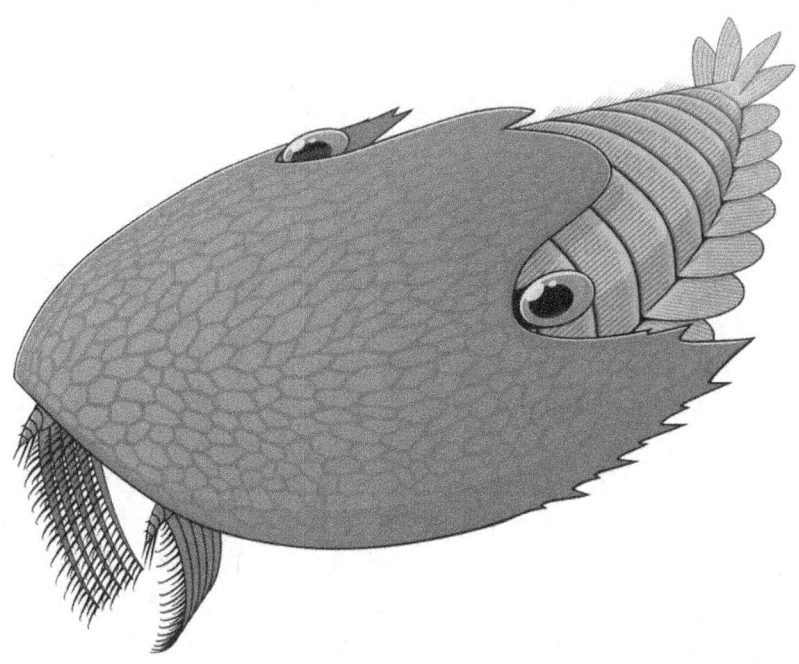

Figure 1.1 *Cambroraster falcatus* reconstruction

I heard a yawn as he stretched. "Yeah, that's really cool!" He then checked his phone. "Two minutes after two."

I nodded. "We made good time."

Leaving Vancouver at 5:00 that morning—or "insanely early," as Seb had put it—we had avoided the traffic congestion of late July in the Lower Mainland. His griping was a token gesture, as we both knew he liked getting a jump on the day, especially when it was the first day of a holiday in the outdoors.

At the eastern edge of British Columbia, in Yoho National Park, is spectacular Takakkaw Falls, Canada's third-highest waterfall. Dropping 833 feet, the thundering water can be heard well before you reach the parking lot at the access road's terminus. As we approached there, a few minutes after listening to the news clip, I said, "What I thought we'd do is claim our campsite, take a quick look at the falls, then backtrack to Field so we can go to the Visitor Center before it

EXCERPT Mysteries of the Origin of Animals — Chapter 1

closes. We can make dinner back at the campsite and take a better look at the falls then."

"Sure," he said. "Is there something in particular you want to get at the Visitor Center?"

"Probably not, but I would like you to see the exhibit they have on the Burgess Shale. Tomorrow, we'll hike up to Yoho Lake and back. The tour to the fossil beds on Wednesday starts at 7:00am, before the Center opens, so this is our opportunity to take a look at some information before that."

He gave a thumbs-up as I parked the Jeep. "Sounds like a plan."

Doing at least one trip a year together had become something of a tradition for us, starting when he was in his early teens. This summer, we'd agreed to do a multi-day backpacking trek in the Canadian Rockies. We also both wanted to see the famed fossil beds that had made this area a UNESCO world heritage site. Discovered in 1909, the Burgess Shale is renowned for the outstanding preservation of its fossils, including soft-bodied creatures and the soft parts of animals with hard body parts, captured for posterity 508 million years ago. Over 60,000 unique fossils have been identified there, including numerous creatures found nowhere else on Earth. To protect them from damage and destruction, the fossil beds could only be visited by guided tour. It was an eleven-hour round trip, hence the early start time.

Having trekked to the campsite area, found our spot, and quickly set up our tent, we headed back to the vehicle and arrived at the Visitor Center as the mid-afternoon sun gave the lush vegetation in the Kicking Horse River valley a particular vibrancy. Rising on all sides above the greens and golds were the dark and light gray hues of towering peaks, the highest still capped with snow.

Inside, we beelined it for the exhibit and spent the next quarter hour going from one display to the next, reading and chatting. A short distance away was a small group listening to a Parks Canada employee relating some history. We caught the tail end:

> At the end of August in 1909, Walcott was riding between Wapta Mountain and Mount Field when he happened to

come across "many interesting fossils," as he wrote in his journal that evening, back at camp with his wife, Helena, and one of their sons, Stuart.

The next day, the three of them returned to the area and found a number of other unusual fossils. Walcott sketched them right there on the site, in his field notebook, then later wrote in his journal: "Out with Helena, Stuart collecting fossils from the Stephen Formation. We found a remarkable group of Phyllopod crustaceans – Took a large number of fine specimens to camp." The next day, they "found a fine group of sponges on slope (in-situ)." These fossils were of animal types never previously seen.

By the end of the 1909 field season, they had spent five days collecting fossils from that area. Even from this brief beginning, Walcott knew they had found something very important. And it was so significant that paleontologists are still exploring it, 110 years later.

Figure 1.2. Charles D. Walcott at the Burgess Shale, c. 1915. Walcott is front left, sitting down.

After the talk ended, the speaker strolled over to us and struck up a conversation. Hearing that we would be on the long tour in two days' time, he began talking about the marvels of the fossil beds. He clearly knew his stuff and was only too glad to help us become familiar with what we would encounter.

As the three of us stood gazing at the image of a beautiful fossilized arthropod called *Marrella splendens*, now extinct, the employee murmured, "Isn't it just amazing what evolution has produced?"

"Evolutionary theory has a tough time taking the credit for the Cambrian explosion," I said.

Seb looked at me quizzically as the employee replied, "What do you mean? You don't think this animal evolved?"

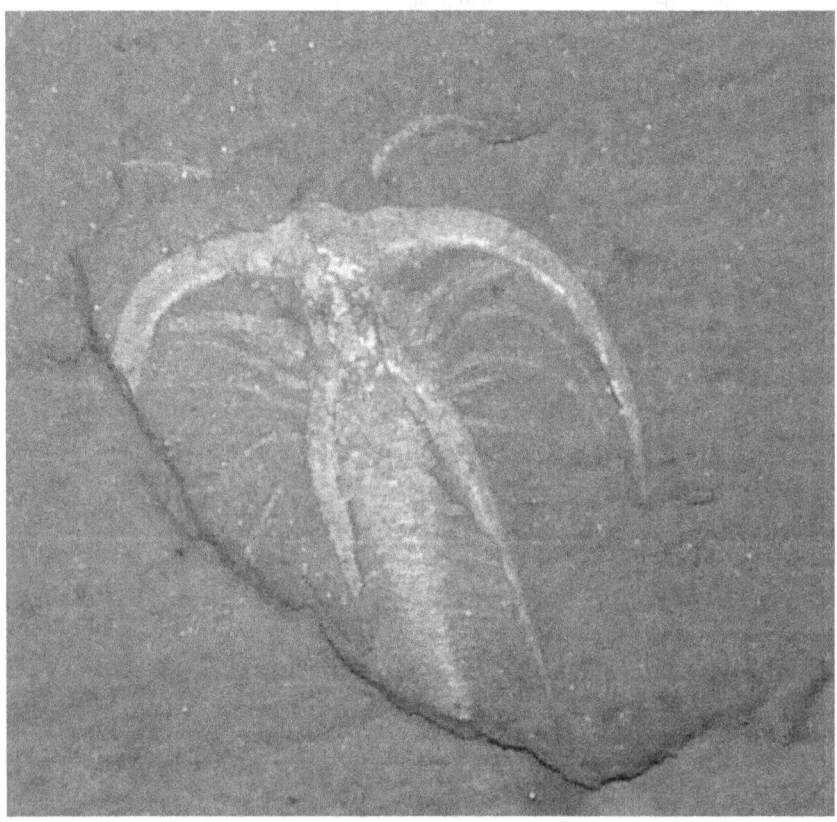

Figure 1.3. Fossil of *Marrella splendens*

"I'm a scientist, and I stick with what the physical evidence tells us about life leading up to the Cambrian events. And when we look at the simplicity and limited range of life forms in the pre-Cambrian fossil record, compared with the incredible complexity of the fossilized forms we see here, it's hard to see how it could have happened through evolution."

Seb interjected, "Well, a billion years is a long, long time."

"I agree. But this life appeared in what's called the Cambrian explosion, most of which happened in just several million years, and the forms we find are diverse and exquisitely complex right from the start." As Richard Dawkins put it: "it is as though [the evolutionary phyla] were just planted there, without any evolutionary history."

At this point, another visitor stopped a short distance from us, clearly hoping to catch the man's attention, so Seb and I stepped back slightly to make room. "It was good to speak with you," I said.

"Yeah, thanks!" added Seb. The employee smiled, nodded a "you're welcome," and turned to the waiting woman.

Within a few more minutes, Seb and I had come to the end of the exhibit and decided to head back to our campsite.

As we turned onto Yoho Valley Road, Seb said, "Dan, what did you mean back there, when the Parks Canada guy mentioned evolution, and you said it couldn't account for the Cambrian explosion?"

I smiled. "It's part of the research I've been doing for my next book, on the origins of animal life."

His eyebrows went up. "Are you trying to combine faith and science to fully understand the origins of life, like you showed me about the first instant of the universe?"

"Yes. I've been working on the appearance of first life, then animals, in particular how complex body plans came about, the development of blood, and the various distinctive categories of animals." Glancing at him, I said, "I'll be happy to talk about it. But we'll need a bit of time. And after traveling for nine-plus hours, I'd rather wait until I've had some food and a good night's sleep."

"Fair enough," said Seb. "Me too, actually. Because if it's anything like our discussions on some of our other trips, I suspect it might take a few days!"

I laughed. He was right. On a couple of previous holidays, we had ended up talking extensively about the ideas in two of my previous books, *The Biblical Clock* and *Mysteries of the First Instant*, which I'd been writing at the time.

"Yes, knowing that you like to go from beginning to end, it'll take some time. We need to discuss the Cambrian before and during our Burgess Shale tour in two days. So I suggest that tomorrow, we talk about how first life came about, which created the cells that Cambrian animals were made of."

"Sounds good to me!"

Deep in the valleys of the Rocky Mountains, dusk comes earlier than you initially expect, as the sun slips behind the great peaks and ridges long before it reaches the true horizon. We decided to walk to the falls while the light levels were good, then eat.

In the Cree language, Takakkaw means "wonderful," and no one gazing up at those white waters cascading between orange-tinged gray cliff faces could possibly disagree. Seb and I stayed until our rumbling stomachs out-competed the thundering flow, then returned to camp for an enjoyable meal and, eventually, a sound sleep.

≈≈≈

To be notified of the book's publication follow Daniel Friedmann here https://www.amazon.com/Daniel-Friedmann/e/B005LKD1Q4/

Or please go to https://www.danielfriedmannbooks.ca/contact/

Endnotes

A NOTE FROM THE AUTHOR

1. Genesis 1:1.

2. Brian Greene, *The Fabric of the Cosmos: Space Time and the Texture of Reality* (New York: Vintage Books, 2005), p. 272.

3. Daniel Friedmann, *The Genesis One Code* (Vancouver: Inspired Books, 2014).

CHAPTER 1

1. "The LEGO® System in Play," https://www.lego.com/en-us/themes/lego-history/articles/lego-system-in-play-60d5efbce6cf46a78794f108b3c19bda

2. We'll look at the Big Bang Theory in more detail in Chapter 2.

3. beyondearthlyskies.blogspot.ca/2013/05/black-holes-on-outskirts.html

4. The same process is used with stars and galaxies. We first understand and classify them. Then we look at stars and galaxies at different distances and therefore ages and figure out how they evolve over time.

5. https://en.wikipedia.org/wiki/Lego#Design

6. Talmud, Pesachim 54a.

7. Talmud, Shabbat 88a; Zohar III:193a, 298b; Rashi on Genesis 1:31.

8. Jerusalem Talmud, Berachot 9a.

9. Midrash Rabbah on Genesis 1:2. See also the Zohar I:134a, Vol. II, 161b.

10. Pirkei Avot 5:26; "Delve in it [the Torah] and continue to delve in it, for everything is in it."

11. Darkness in Torah is not an absence of light but a separate ex nihilo creation. (Isaiah 45:7)

12. "Torah and Mathematics: The Story of Π – Part 1," www.inner.org/torah_and_science/mathematics/story-of-pi-1.php

"Torah and Mathematics: The Story of Π – Part 2,"
www.inner.org/torah_and_science/mathematics/story-of-pi-2.php

13 I Kings 7:23.

CHAPTER 2

1. J. J. O'Connor and E. F. Robertson, "Aleksandr Aleksandrovich Friedmann," School of Mathematics and Statistics, University of St. Andrews, 1997. www-history.mcs.st-andrews.ac.uk/Biographies/Friedmann.html

2. Judd D. Rogers et al., "An Absorption Profile Centred at 78 Megahertz in the Sky-Averaged Spectrum," *Nature*, February 28, 2018. https://arxiv.org/abs/1810.05912

3. Planck Collaboration, "Planck 2015 Results. XIII. Cosmological Parameters," *Astronomy & Astrophysics*, 594 (2015): A13. doi:10.1051/0004-6361/201525830

4. (i) E. Harrison, *Cosmology—The Science of the Universe*, 2nd ed. (Cambridge: Cambridge University Press, 2003).
   (ii) B. Ryden, *Introduction to Cosmology*, 2nd ed. (Cambridge: Cambridge University Press, 2017).

CHAPTER 3

1. "Black Holes on the Outskirts," Beyond Earthly Skies. http://beyondearthlyskies.blogspot.ca/2013/05/black-holes-on-outskirts.html

2. "Structure of the Universe 2," Sun.org. http://www.sun.org/images/structure-of-the-universe-2

3. (i) A. G. Reiss et al., "Observational Evidence from Supernovae for an Accelerating Universe and a Cosmological Constant," *Astronomical Journal*, 116.3 (1998): 1009–1038. doi:10.1086/300499
   (ii) A. G Riess et al., "A 2.4% Determination of the Local Value of the Hubble Constant," *Astrophysical Journal*, 826.1 (2016). https://iopscience.iop.org/article/10.3847/0004-637X/826/1/56

4. A. G. Reiss and Mario Livio, "The Puzzle of Dark Energy," *Scientific American*, Wonders of the Cosmos, special edition, 2017.

5. Reiss and Livio, 2017.

6. Planck Collaboration, "Planck 2015 Results. XIII. Cosmological Parameters," *Astronomy & Astrophysics* 594 (2015): A13. doi:10.1051/0004-6361/201525830

CHAPTER 4

1. (i) Mark Whittle, Cosmology: *The History and Nature of Our Universe*, Course Guidebook (Chantilly, VA: The Teaching Company, 2008), pp. 169–173.
   (ii) Paul J. Steinhardt, "The Inflation Debate—Is the Theory at the Heart of Modern Cosmology Deeply Flawed?" *Scientific American*, April 2011, pp. 38–43.
2. For example, one way to measure the curvature of the universe is to examine the cosmic microwave background (CMB) radiation. The CMB is electromagnetic radiation that fills the universe, left over from an early stage in its history. The temperature of this radiation is almost the same at all points in the sky, but there is a slight variation (around one part in 100,000) between the temperature received from the different points in the sky. The angular scale of these fluctuations (the typical angle between a hot patch and a cold patch in the sky) depends on the curvature of the universe. Thus, by measuring these fluctuations we determine that the universe is flat.

CHAPTER 5

1. Alan Guth, "Stumbling to Inflation" [Audio file]. The Story Collider: True Personal Stories about Science. November 27, 2015. https://www.storycollider.org/stories/2015/12/13/alan-guth-stumbling-to-inflation
2. Brian Greene, *The Fabric of the Cosmos: Space, Time, and the Texture of Reality* (New York: Vintage Books, 2005), pp. 322–323.
3. Brian Greene, *The Fabric of the Cosmos: Space, Time, and the Texture of Reality* (New York: Vintage Books, 2005), pp. 322–323.
4. Niayesh Afshordi, Robert B. Mann, and Razieh Pourhasan, "The Black Hole at the Beginning of Time," *Scientific American*, December 2015. https://www.scientificamerican.com/article/the-black-hole-at-the-beginning-of-time/
5. Anna Ijjas, "What If There Was No Big Bang and We Live in an Ever-Cycling Universe?" *New Scientist*, August 14, 2019. https://www.newscientist.com/article/mg24332430-800-what-if-there-

was-no-big-bang-and-we-live-in-an-ever-cycling-universe/#ixzz5z5ofVWST

6 Steven Weinberg, "Physics: What We Do and Don't Know." *New York Review of Books*, November 20, 2007. If the multiverse exists, "the hope of finding a rational explanation for the precise values of quark masses and other constants of the standard model of particle physics that we observe in our Big Bang is doomed, for their values would be an accident of the particular part of the multiverse in which we live."

7 Joshua Howgego, "10 Mysteries of the Universe: How Did It All Begin?" *New Scientist*, September 19, 2018. https://www.newscientist.com/article/mg23931960-200-10-mysteries-of-the-universe-how-did-it-all-begin/

CHAPTER 6

1 CERN, CERN and the Rise of the Standard Model [video], 2014. https://home.cern/science/physics/standard-model

2 See the National Institute of Standards and Technology's webpage "Unit of Time (Second)" at physics.nist.gov/cuu/Units/second.html

3 The exact value is 1/299,792,458; see "Unit of Length (Meter)" at physics.nist.gov/cuu/Units/meter.html

CHAPTER 7

1 Gershom Scholem, *Kabbalah* (New York: Times Books, 1978), p. 3.

2 Yitzchak Ginsburgh, *What You Need to Know About Kabbalah* (Jerusalem: Gal Einai Publications, 2006), pp. 17–19.

3 Rabbi Menachem Mendel of Lubavitch, *Derech Mitzvosecha: A Mystical Perspective on the Commandments*, Volume II: Mitzvas Haamanas Elokus (New York: Sichos In English, 2007), p. 171.

4 The Ari taught orally, writing down little of his own teachings. His students (and students' students) documented them, particularly his closest student, Chayim Vital (1543–1620).

5 The numerical value, or gematria, of tohu in Hebrew is 411, the same as the numerical value for the phrase yesh me ayin, or "something from nothing." The average value of the three letters of tohu, 411/3, is 137—the fine structure constant; see Rabbi Yitzchak Ginsburgh, *913: The Secret Wisdom of Genesis* (Jerusalem: Gal Einai, 2015), p. 135.

Endnotes

6 Ramban (Nachmanides), translated and annotated by Charles B. Chavel, *Commentary on the Torah, Genesis* (New York: Shilo Publishing House, 1971), p. 25.

7 Meir Zlotowitz, Bereishis, *Genesis: A New Translation with a Commentary Anthologized from Talmudic Midrashic and Rabbinic Sources* (New York: Mesorah Publications, 1977), p. 732, on Genesis 1:3.

8 Ramban on Genesis 1:1.

9 Ramban on Genesis 1:1.

10 There are two more instances of bara in the first chapter of Genesis, but they relate to the creation of complex animals and humans, not the universe.

11 "And God said, 'Let us make man in our image, after our likeness.'" Genesis 1:26.

12 Meir Zlotowitz, Bereishis, Genesis, p. 70, on Genesis 1:26.

13 This concept is contained in Exodus 10:2, just before the eighth plague: "and so that you may relate in the ears of your son and your son's son that I made a mockery of Egypt and my signs that I placed among them—that you may know that I am God." The commentators elaborate on this verse to explain that God guides history, "that the Exodus was a seminal event in world history because it demonstrated God's mastery over nature. Thus, it became the textbook lesson for humanity that God is not an aloof Creator, but the Master of the universe day by day and event by event. This verse encapsulates that concept, for it tells Israel that the miracles of the Exodus were to teach them for all generations that God can toy with the most powerful kingdoms, and that this creates the perception that he is YHWH, the name that denotes his eternity, because its letters comprise the word 'He was, He is, He will be'" (Meir Zlotowitz, Bereishis, Genesis, p. 341 on Exodus 10:2).

14 David Schulman, *The Sefirot: Ten Emanations of Divine Power* (London: Jason Aronson Inc., 1996), Introduction.

15 Daniel Friedmann, *Roadmap to the End of Days* (Vancouver: Inspired Books, 2017), Chapter 4.

16 Leviticus 25:3–5.

CHAPTER 8

1. Details about the manuscript's author and intriguing history can be found in Chapters 1 and 2 of Daniel Friedmann and Dania Sheldon, *The Biblical Clock: The Untold Secrets Linking the Universe and Humanity with God's Plan* (Vancouver: Inspired Books, 2019).

2. Selections from *Otzar HaChaim*, 86b–87b, as presented in Aryeh Kaplan, Yaakov Elman, and Israel Lipschutz Ktav, *Immortality, Resurrection and the Age of the Universe: A Kabbalistic View* (Brooklyn, NY: Ktav Publishing House, 1993), pp. 13–14. Kabbalists believed they had derived the secret meaning of various biblical and Talmudic sources that compared the cosmic timeline to the sabbatical cycles of the land—each seven years but with seven cycles required for forty-nine years to the Jubilee—and thereby had discovered the concept of cosmic cycles. There would be, they claimed, seven such cycles, each 7,000 years long, giving a total timeline for the world and universe of 49,000 years. For a longer explanation, see Daniel Friedmann, *The Biblical Clock*, Chapter 2.

3. Adam was completed in hour five of the sixth day. He sinned in hour ten and was expelled by the end of the twelfth hour, or 6pm on the sixth day, Friday. This marked the beginning of the calendar since creation. From Talmud Sanhedrin 38b.

4. The sages of the late Second Temple Period and the century after its destruction calculated the date of Creation. They did so by basing their work upon the Bible's account of lifetimes and kingdoms, thereby determining the period of time from Creation to a known date, in this case, the destruction of the Second Temple in 70 CE. Many rabbis attempted this task, but the method attributed to Rabbi Yossi ben Halafta, a second-century CE sage, gained popularity. He calculated birth "from nothing" to be 3761 BCE. www.myjewishlearning.com/article/counting-the-years/

5. See en.wikipedia.org/wiki/Dating_creation

6. Daniel Friedmann and Dania Sheldon, *The Biblical Clock: The Untold Secrets Linking the Universe and Humanity with God's Plan* (Vancouver: Inspired Books, 2019), Chapter 2.

7. Daniel Friedmann, *The Genesis One Code* (USA: Inspired Books, 2014), Chapters 4–5.

8   Daniel Friedmann, *The Genesis One Code* (USA: Inspired Books, 2014), Chapter 8; Daniel Friedmann, *The Broken Gift* (USA: Inspired Books, 2013).

9   Zohar Vayera 119a, Ramban on Genesis 2:3 maintains that the seven days of Creation correspond to the seven millennia of the existence of natural creation. The tradition teaches that the seventh day of the week, Shabbat or the day of rest, corresponds to the Great Shabbat, the seventh millennium (years 6000–7000), the age of universal rest.

10   Daniel Friedmann, *Roadmap to the End of Days* (USA: Inspired Books, 2017), Chapter 1.

11   spaceplace.nasa.gov/sun-age/en/

12   In this approach, God created a theoretical timeline of all of existence, starting from the Big Bang and spanning billions of years. In this theoretical timeline, stars were born and died, dinosaurs roamed the Earth, and different human-like creatures came and went. Then, at a very specific point in this timeline, God took a snapshot of the entire universe, exactly as it would look at that moment, and that's what He created. In other words, God brought a world billions of years old into existence fewer than 6,000 years ago. For more see Daniel Friedmann and Dania Sheldon, *The Biblical Clock: The Untold Secrets Linking the Universe and Humanity with God's Plan* (Vancouver: Inspired Books, 2019), Chapter 6.

13   Michael Friedlander, *Pirkê de Rabbi Eliezer* [part of the Oral Law]. (Illinois: Varda Books, 2004), Chapter 7.

14   Daniel Friedmann, *The Genesis One Code* (USA: Inspired Books, 2014), Chapters 6 and 7.

15   Planck Collaboration, "Planck 2015 Results. XIII. Cosmological Parameters," *Astronomy & Astrophysics,* 594 (2015): A13. doi:10.1051/0004-6361/201525830

16   Eduardo F. del Peloso et al., "The Age of the Galactic Thin Disk from Th/Eu Nucleocosmochronology: Extended Sample," *Proceedings of the International Astronomical Union*, 1 (December 23, 2005), pp. 485–486.

17   Less than 5 billion years ago, the gas mass that was to be the sun detached from a bigger mass. That predetermined the mass of the sun. (Dunham et al,. "The Evolution of Protostars: Insights from Ten Years of Infrared Surveys with Spitzer and Herschel" https://muse.jhu.edu/chapter/1386885). The gas cloud then contracted under its own gravity and began to glow, finally igniting into

nuclear fusion 4.57 billion years ago (James N. Connelly et al., "The Absolute Chronology and Thermal Processing of Solids in the Solar Protoplanetary Disk," *Science*, 338.6107 (November 2012): 651–655. doi:10.1126/science.1226919

18   Thorsten Klein et al., "Hf–W Chronometry of Lunar Metals and the Age and Early Differentiation of the Moon," *Science Magazine*, 310/5754 (2005): 1671–1674.

19   Daniel Friedmann, *The Genesis One Code* (USA: Inspired Books, 2014), Chapter 7.

20   Daniel Friedmann, *The Genesis One Code* (USA: Inspired Books, 2014), Chapter 7.

21   Daniel Friedmann, *The Genesis One Code* (USA: Inspired Books, 2014), Chapter 7.

22   The scientific timeline is billions of years. The biblical timeline is six Creation days. In the science view things came about naturally and required a long evolutionary period. In the biblical teaching things were made ready to use, in final form, not requiring time to develop. In this view God brought an "old" universe into existence. See note 12 above.

CHAPTER 9

1   Rabbi Menachem Mendel was known then as the Rebbe—which is the title for the spiritual leader in the Hasidic movement—but today he is known as the Tzemach Tzedek (the title of a very large compendium of Jewish law that he authored) or simply by his name.

2   Rabbi Menachem Mendel of Lubavitch, *Derech Mitzvosecha—A Mystical Perspective on the Commandments* (Brooklyn, NY: Sichos in English, 2007), Chapter 6; available at http://www.hebrewbooks.org/15419 (English) and http://hebrewbooks.org/16082 (Hebrew).

3   "The truth is that even after the creation of worlds, there is nothing outside of His Essence. Nevertheless, so it arose in His will that the world should be an entity that appears independent from Him, and that it should be limited." Rabbi Yosef Yitzchak Schneersohn of Lubavitch, *Creation and Redemption* (Brooklyn, NY: Kehot Publication Society, 2007), pp. 48–49, footnotes 51–55.

4   Rabbi Menachem Mendel of Lubavitch, *Derech Mitzvosecha—A Mystical Perspective on the Commandments* (Brooklyn, NY: Sichos in English, 2007),

Chapter 6; available at http://www.hebrewbooks.org/15419 (English) and http://hebrewbooks.org/16082 (Hebrew).

5   Rabbi Schneur Zalman of Liadi, *Likutei Amarim, Part Two: Shaar Hayichud Vehaemuna*, bilingual edition (Brooklyn, NY: Kehot Publication Society, 1993), Chapter 7; available at http://www.chabad.org/library/tanya/tanya_cdo/aid/1029162/jewish/Shaar-Hayichud-Vehaemuna.htm

6   Tikkunin, Tikkun 57. *Tikkunim* is a main text of the Kabbalah. It is a separate appendix to the Zohar, consisting of seventy commentaries on the opening word of the Torah.

7   Rabbi Menachem Mendel of Lubavitch, *Derech Mitzvosecha—A Mystical Perspective on the Commandments* (Brooklyn, NY: Sichos in English, 2007), Chapter 6; available at http://www.hebrewbooks.org/15419 (English) and http://hebrewbooks.org/16082 (Hebrew).

8   Talmud Chagigah 12a.

9   Ramban (Nachmanides), translated and annotated by Charles B. Chavel, *Commentary on the Torah, Genesis* (New York: Shilo Publishing House, 1971), p. 25.

10 Term coined by Ramban; see Ramban (Nachmanides), translated and annotated by Charles B. Chavel, *Commentary on the Torah, Genesis* (New York: Shilo Publishing House, 1971), p. 25.

11 Rabbi Schneur Zalman of Liadi, *Likutei Amarim, Part Two: Shaar Hayichud Vehaemuna*, bilingual edition (Brooklyn, NY: Kehot Publication Society, 1993), Chapter 7; available at http://www.chabad.org/library/tanya/tanya_cdo/aid/1029162/jewish/Shaar-Hayichud-Vehaemuna.htm

12 Genesis 1:27, 2:7. Rashi's commentary on "and breathed into his nostrils" (Genesis 2:7).

13 "[We] can never know with any degree of certainty how the universe came about, if scientific knowledge is to be the only source of our knowledge." Letter from Rebbe Rabbi Menachem M. Schneerson dated 17th of Cheshvan, 5723, Brooklyn, NY. The Letter & the Spirit: Letters by the Lubavitcher Rebbe Rabbi Menachem M. Schneerson, selected and arranged by his personal secretary, Rabbi Dr. Nissan Mindel OBM, Volume II, Nissan Mindel Publications 5773, 2013, Brooklyn, NY, p. 236.

14   Jacob Immanuel Schochet, *Mystical Concepts in Chassidism: An Introduction to Kabbalistic Concepts and Doctrines* (Brooklyn, NY: Kehot Publication Society, 1979), Chapter 2; available at http://www.hebrewbooks.org/15600

15   "God conveys His influence through the ten sefirot. The light in them is the same; how they differ is due to the vessels that the light goes into, called the keli. The general principle that applies is that the influence changes according to the nature of the keli through which the light passes. On the level identified with the *Sefiros* of *Tohu*, the *Sefiros* exist as ten points of light. And on the level identified with the sefirot of *Tikkun* each one subdivides into ten, allowing for interrelation between the *Sefiros* and thus establishing a broad base of stability for the *Sefiros* as they emerge into distinct entities." From Rabbi Menachem Mendel of Lubavitch, *Derech Mitzvosecha—A Mystical Perspective on the Commandments* (Brooklyn, NY: Sichos in English, 2007), Chapter 4; available at http://www.hebrewbooks.org/15419 (English) and http://hebrewbooks.org/16082 (Hebrew).

16   "In this context, R. Isaac Luria reads Genesis 1:2 as referring to the World of Tohu, and Genesis 1:3 ('Let there be light, and there was light') as referring to the World of Tikun; *Etz Chayim* 8:1." From Jacob Immanuel Schochet, *Mystical Concepts in Chassidism: An Introduction to Kabbalistic Concepts and Doctrines* (Brooklyn, NY: Kehot Publication Society, 1979), p. 143; available at www.hebrewbooks.org/15600

17   God conveys his influence through the ten sefirot. The way they differ is due to the vessels that the light goes into—the keli. "The general principle that applies is that the influence changes according to the nature of the keli through which the light passes. ... On the level of *Akudim* (identified with the level of *Atik Yamin*), the *Sefiros* exist as ten lights bound (*akudim*) in one *k'li*. On the level of *Nekudim* (identified with the *Sefiros* of *Tohu*), the *Sefiros* exist as ten points (*nekudim*) of light. And on the level of *Verudim* (identified with the gestalt of *Tikkun*) each one subdivides into ten, allowing for interrelation (*verudim*) between the *Sefiros* and thus establishing a broad base of stability for the *Sefiros* as they emerge into distinct entities in the realm of *Atzilus*." From Rabbi Menachem Mendel of Lubavitch, *Derech Mitzvosecha—A Mystical Perspective on the Commandments* (Brooklyn, NY: Sichos in English, 2007), Chapter 4; available at http://www.hebrewbooks.org/15419 (English) and http://hebrewbooks.org/16082 (Hebrew).

18   "In this context R. Isaac Luria reads Genesis 1:2 as referring to the World of *Tohu*, and Genesis 1:3 ('Let there be light, and there was

light') as referring to the World of *Tikun; Etz Chayim* 8: 1." From Jacob Immanuel Schochet, *Mystical Concepts in Chassidism: An Introduction to Kabbalistic Concepts and Doctrines* (Brooklyn, NY: Kehot Publication Society, 1979), p. 143; available at www.hebrewbooks.org/15600

19 (i) Moshe Miller, "Shattered Vessels," www.chabad.org/kabbalah/article_cdo/aid/380568/jewish/Shattered-Vessels.htm
    (ii) Gershom Scholem, *Kabbalah* (New York: Times Books, 1978), pp. 135–144.

20 (i) Moshe Miller, "Shattered Vessels," www.chabad.org/kabbalah/article_cdo/aid/380568/jewish/Shattered-Vessels.htm
    (ii) Gershom Scholem, *Kabbalah* (New York: Times Books, 1978), pp. 135–144.
    (iii) Yitzchak Ginsburgh, *Lectures on Torah and Modern Physics* (Jerusalem: Gal Einai Publications, 2013), p. 104.

21 (i) "He created substance from tohu, and made that which was nothing something"; Rabbi Aryeh Kaplan, *Sefer Yetzirah—The Book of Creation: In Theory and Practice* (San Francisco: Weiser Books, 1997), 2:6.
    (ii) Ramban (Nachmanides), translated and annotated by Charles B. Chavel, *Commentary on the Torah, Genesis* (New York: Shilo Publishing House, 1971), pp. 22–27: "In the beginning God created the ... earth means that he brought forth its matter from nothing. ... earth includes all four elements. After having said that with one commandment God created at first ... the earth, Scripture returns and explains that the earth after this creation was tohu, that is, matter without substance. It became tohu when he clothed it with form. Then Scripture explains that in this form was included the four elements. ... [N]ow it is already known that the four elements fill up the whole space with matter."
    (iii) Rambam (Nachmanides), *Guide for the Perplexed*, I, 72, trans. Michael Friedlander: "At the briefest instant following creation all the matter of the universe was concentrated in a very small place. The matter at this time was very thin, so intangible, that it did not have real substance. It did have, however, a potential to gain substance and form and to become tangible matter. From the initial concentration of this intangible substance in its minute location. This initially thin non-corporeal substance took on the tangible aspects of matter as we know it. From this initial act of creation,

from this ethereally thin pseudosubstance, everything that has existed, or will ever exist, was, is, and will be formed."

22   Jacob Immanuel Schochet, *Mystical Concepts in Chassidism—An Introduction to Kabbalistic Concepts and Doctrines* (Brooklyn, NY: Kehot Publication Society, 1979), p. 136. Available at www.hebrewbooks.org One can look at the light as letters and the vessels as words—they contain the letters and generate a concept, i.e., the meaning of the word. By shattering the vessels, the words were broken apart into the constituent letters. The letters are then free to form an infinity of combinations, i.e., other "words." The letters at the lowest level of existence correspond to the elementary particles that make up matter.

23   The vessels of tohu were situated one above the other in a single line, indicating that they acted as independent entities. Thus, each vessel of tohu "existed as an autonomous fiefdom, so to speak, independent of, and even in opposition to, the others." In addition, "the vessels themselves were in a state of immaturity and were therefore unable to contain the intense light flooding them." In contrast, the vessels of the world of tikkun were arranged in harmonious interrelated triads; see Figure 12.1; Miller, "Shattered Vessels," https://www.chabad.org/kabbalah/article_cdo/aid/380568/jewish/Shattered-Vessels.htm

24   Ramban (Nachmanides), translated and annotated by Charles B. Chavel, *Commentary on the Torah, Genesis* (New York, NY: Shilo Publishing House, 1971), pp. 8–10.

25   In the Torah, "nothing" means nothing physical—no time, no space, no forces of nature, no elementary particles. According to the Torah, in the beginning, God created from nothing physical. In philosophy as well, "nothing" means nothing physical. In science, "nothing" most of the time actually still means something—typically, at least gravity and space. Often, "nothing" is referred to as a quantum vacuum. The quantum vacuum is that very early state of the universe in the first fraction of a second when the universe was so hot and dense that physical particles could not exist. However, according to the present-day understanding of the "vacuum state" or the quantum vacuum, it was and is by no means a simple empty space. Quantum mechanics holds that a vacuum state is not truly empty but instead contains fleeting electromagnetic waves and particles that pop into and out of existence. In the quantum vacuum at the beginning of the universe, time, space, the laws of physics, and particles all existed. However, the

particles did not endure as physical entities because at such a high temperature, as soon as they appeared, they turned back into energy—they were "virtual" particles. Due to the apparent absence of physical particles, it seemed like there was nothing, but in reality, everything needed to build the universe existed. As the universe expanded and cooled, the particles came into being and remained; eventually, the stars and galaxies formed. Notably, the Torah talks about and explains the quantum vacuum state in detail as the world of tohu. This is alluded to in Genesis 1:2: "Now the earth was astonishingly empty [tohu va vohu], and darkness was on the face of the deep."

26 All the steps described herein are in the spiritual strata, not the physical. However, our physical world is the culmination of contractions from the spiritual—i.e., "chochma of azilut serves as the source of mortal intellect though a myriad of chain-like descents"; Rabbi Menachem Mendel of Lubavitch, *Derech Mitzvosecha—A Mystical Perspective on the Commandments* (Brooklyn, NY: Sichos in English, 2007), p.185; available at http://www.hebrewbooks.org/15419 (English) and http://hebrewbooks.org/16082 (Hebrew). As noted in the student textbook *The Kabbalah of Time*, the "physical is a reflection of the spiritual template" (Brooklyn, NY: The Rohr Jewish Learning Institute, n.d.), lesson 2. Thus, later in the book, we take the description of the physical to be similar to that of the spiritual when comparing it to what science has found.

27 Lubavitcher Rebbe, adapted by Yanki Tauber, *Inside Time: A Chassidic Perspective on the Jewish Calendar* (Brooklyn, NY: Meaningful Life Center, 2015), volume 1, p. 7.

28 "The truth is that even after the creation of worlds, there is nothing outside of His Essence. Nevertheless, so it arose in His will that the world should be an entity that appears independent from Him, and that it should be limited"; Rabbi Yosef Yitzchak Schneersohn of Lubavitch, *Creation and Redemption* (Brooklyn, NY: Kehot Publication Society, 2007), pp. 48–49, footnotes 51–55.

29 Lubavitcher Rebbe, adapted by Yanki Tauber, *Inside Time: A Chassidic Perspective on the Jewish Calendar* (Brooklyn, NY: Meaningful Life Center, 2015), volume 1, p. 9.

30 Lubavitcher Rebbe, adapted by Yanki Tauber, *Inside Time: A Chassidic Perspective on the Jewish Calendar* (Brooklyn, NY: Meaningful Life Center, 2015), volume 1, p. 7.

31 Rabbi Menachem Mendel of Lubavitch, *Derech Mitzvosecha—A Mystical Perspective on the Commandments* (Brooklyn, NY: Sichos in English, 2007), Chapter 11; available at http://www.hebrewbooks.org/15419 (English) and http://hebrewbooks.org/16082 (Hebrew).

32 Rabbi Menachem Mendel of Lubavitch, *Derech Mitzvosecha—A Mystical Perspective on the Commandments* (Brooklyn, NY: Sichos in English, 2007), Chapter 12; available at http://www.hebrewbooks.org/15419 (English) and http://hebrewbooks.org/16082 (Hebrew).

33 Daniel Friedmann, *Roadmap to the End of Days* (USA: Inspired Books, 2017), Chapter 1.

34 Rabbi Menachem Mendel of Lubavitch, *Derech Mitzvosecha—A Mystical Perspective on the Commandments* (Brooklyn, NY: Sichos in English, 2007), Chapter 11; available at http://www.hebrewbooks.org/15419 (English) and http://hebrewbooks.org/16082 (Hebrew).

35 Jacob Immanuel Schochet, *Mystical Concepts in Chassidism: An Introduction to Kabbalistic Concepts and Doctrines* (Brooklyn, NY: Kehot Publication Society, 1979), p. 136; available at www.hebrewbooks.org/15600

36 Yitzchak Ginsburgh, *Lectures on Torah and Modern Physics* (Jerusalem: Gal Einai, 2013), p. 104. "General relativity is the world of tikun while quantum mechanics is an example of the world of tohu."

37 Quantum mechanics has shown that the microscopic world is not bound by space and time. The following is an example with respect to space. Interferometry experiments show that when an electron has two ways to travel to a destination, a single electron travels both ways. If we try to measure which way it goes, we do so and know which way it went; but the end result is different than when we don't know and it goes "both ways." There is no explanation for this behavior. The electron is simply not confined to space as we know it. Richard Feynman has shown mathematically that if an electron is going from point a to point b, it "travels along every conceivable path." The calculations in quantum mechanics, which is the most accurate theory of physics, actually are performed by calculating each path the electron can take and adding it to all the other calculated paths to obtain the total answer for the electron going from point a to b (i.e., the probability that it gets from a to b); Benjamin Schumacher, *Quantum Mechanics: The Physics of the Microscopic World*. Course Guidebook (Chantilly, VA: The Teaching Company, 2009), Chapter 17. Here is an example with respect to time. Two particles in a quantum system may be entangled. That is, their states are connected. For instance, two

particles can have a spin of zero, meaning that each has the opposite spin half of the time—i.e., when we measure them, they are always opposite, even after we switch the spin of one. That the second particle switches its spin automatically after we cause the spin of the first to switch holds true even if the particles are very far apart, leading to what Einstein called spooky action at a distance. The only way to deal with this is either to abandon the notion that particles have a spin before we measure them, or to postulate that the particles in the universe are all connected by a web of instantaneous communication links—in other words, they are beyond time and space. Schumacher, *Quantum Mechanics*, Chapters 15 and 16.

38 Brian Greene, *The Fabric of the Cosmos: Space, Time and the Texture of Reality* (New York: Vintage Books, 2005), pp. 322–323.

39 (i) Jacob Immanuel Schochet, *Mystical Concepts in Chassidism: An Introduction to Kabbalistic Concepts and Doctrines* (Brooklyn, NY: Kehot Publication Society, 1979), Chapters 7–9; available at http://www.hebrewbooks.org/15600;
(ii) Gershom Scholem, *Kabbalah* (New York: Times Books, 1978), pp. 135–144.

40 Unchanging strictly only from 5,781 years ago, as creations could evolve during the six days—equivalent to 13.8 billion years—and were only strictly fixed at the end of the six days.

41 (i) "All composite particles in nature decay into proton[s] (which [are] made of quarks that are not found alone, only as protons of neutrons), photon, electron, and electron type neutrino[s] which are the stable particles of nature"; Suresh Emre, "Why Elementary Particles Decay," *Renaissance Universal*, sureshemre.wordpress.com/2013/11/17/why-elementary-particles-decay
(ii) As will be explained later, the elementary particles according to Kabbalah come in three groups (as science has found). Only the lowest group (up, down quarks, electrons, and neutrinos) are permanent and exist forever. The higher groups do not. All groups are perfect and have constant properties, such as spin. Finally, massless particles also exist forever if allowed, since they do not experience time.

42 We will see in Part 2 of the book that only the lowest-generation particles live forever; the higher generations have a limited life. But everything we see is made only of the lower-level particles.

43 Malbin, Genesis 1:2.

44 Moshe Chaim Luzzatto (RaMCHaL), *Klach Pitchei Chochma (138 Openings to Wisdom)*, Petach 18, azamra.org/Kabbalah/Openings/018.htm; for the home page of the whole book, see azamra.org/Openings.shtml

45 Moshe Chaim Luzzatto (RaMCHaL), *Klach Pitchei Chochma (138 Openings to Wisdom)*, Petach 19, azamra.org/Kabbalah/Openings/019.htm

46 Rabbi Yitzchak Luria (the Arizal), *Etz Hayim (Tree of Life)*, 5:3.

47 Rabbi Schneur Zalman of Liadi, *Likutei Amarim, Part Two: Shaar Hayichud Vehaemuna*, bilingual edition (Brooklyn, NY: Kehot Publication Society, 1993), Chapter 2; available at www.chabad.org/library/tanya/tanya_cdo/aid/1029162/jewish/Shaar-Hayichud-Vehaemuna.htm

CHAPTER 10

1 Rabbi Menachem Mendel of Lubavitch, *Derech Mitzvosecha—A Mystical Perspective on the Commandments* (Brooklyn, NY: Sichos in English, 2007), Chapter 6; available at http://www.hebrewbooks.org/15419 (English) and http://hebrewbooks.org/16082 (Hebrew).

2 The Torah contains various units of measurement. The smallest measurement of length is "the length of a barleycorn," specified in today's units as 10–11 mm; hence, "small" is this order of measurement. For volume, the smallest unit is equivalent to about an egg. Shlomo Ganzfried, Kitzur Shulchan Aruch Code of Jewish Law (New York: Artscroll, 2009), Appendix A.

3 Babylonian Talmud Chagigah 12a.

4 Babylonian Talmud Chagigah 12a.

5 Hirsch, Genesis 2.1, translates vayehulu as "they were brought to their conclusion."

6 Based on Psalm 148:6: "And He set them up to eternity, yea forever, He issued a decree, which will not change." Rambam explains (*Guide for the Perplexed*, translated by M. Friedländer, 2nd edition (London: Routledge & Kegan Paul, 1904), Part 2, Chapter 30) that this rule became operative after the completion of the six days of creation. Before that time, the laws governing the universe could have been quite different from those we are familiar with today. From Rabbi Moshe

Meiselman, *Torah, Chazal and Science* (Israel: Israel Bookshop Publications, 2013), p. 409.

7 There are two methods for measuring the expansion of the universe. One uses the cosmic microwave background—the afterglow of the Big Bang's primordial fireball from when the universe was only 380,000 years old. This method clearly employs light from billions of years ago. The other uses special stars and is the one described herein.

8 (i) First, they measure the distances to a class of pulsating stars called Cepheid Variables, employing a basic tool of geometry called parallax. This is the same ways our eyes see depth—by looking from two slightly different vantage points at the same object, they measure how far away it is. Once astronomers calibrate the Cepheids' true brightness, they can use them as cosmic yardsticks to measure distances to galaxies much farther away than they can with the parallax technique, eventually extending the process to stars billions of light years away. Once the distance is known, it is matched with how the light from the supernovae is stretched to longer wavelengths by the expansion of space. This gives the stretch as a function of distance, or back in time. See NASA, "Three Steps to Measuring the Hubble Constant," June 2, 2016. www.nasa.gov/image-feature/goddard/2016/three-steps-to-measuring-the-hubble-constant/

(ii) This method yields an expansion rate about 9% higher than the expansion rate measured from the CMB. See Richard Panek, "How a Dispute over a Single Number Became a Cosmological Crisis," *Scientific American*, March 1, 2020. https://www.scientificamerican.com/article/how-a-dispute-over-a-single-number-became-a-cosmological-crisis/

9 Latest measurements use stars at least 100,000 years away; see NASA, "Mystery of the Universe's Expansion Rate Widens with New Hubble Data," April 25, 2019. www.nasa.gov/feature/goddard/2019/mystery-of-the-universe-s-expansion-rate-widens-with-new-hubble-data

10 Aryeh Kaplan, *Sefer Yetzirah—The Book of Creation: In Theory and Practice* (San Francisco: Weiser Books, 1997), 6:1.

11 (i) Aryeh Kaplan, *Sefer Yetzirah—The Book of Creation: In Theory and Practice* (San Francisco: Weiser Books, 1997), pp. 231–239.

(ii) Matter in the universe is distributed in a cosmic web, with a filament of matter typically connecting neighboring galaxy pairs. Scientifically we still don't fully understand the rotation, if any, of structures larger than galaxies. However, recent simulation work has shown that matter generally rotates substantially around intergalactic filaments. See: Xia, Qianli & Neyrinck, Mark & Cai, Yan-Chuan & Aragon-Calvo, Miguel. (2020). Intergalactic filaments spin, arXiv:2006.02418 [astro-ph.CO]

12. (i) Anthony Challinor, "CMB Anisotropy Science: A Review," *Proceedings of the International Astronomical Union*, 8 (2012): 42–52. doi:10.1017/S1743921312016663
    (ii) C. J. Copi, D. Huterer, D. J. Schwarz, and G. D. Starkman, "On the Large-Angle Anomalies of the Microwave Sky," *Monthly Notices of the Royal Astronomical Society*, 367 (2006): 79–102.
    (iii) Antonio Mariano and Leandros Perivolaropoulos, "CMB Maximum Temperature Asymmetry Axis: Alignment with Other Cosmic Asymmetries," *Physical Review D*, 87.4 (2013). doi:10.1103/PhysRevD.87.043511

13. (i) Jacob Aron, "Planck Shows Almost Perfect Cosmos – Plus Axis of Evil," *New Scientist*, March 21, 2013. www.newscientist.com/article/dn23301-planck-shows-almost-perfect-cosmos-plus-axis-of-evil/
    (ii) ESA, "Planck Finds No New Evidence for Cosmic Anomalies," June 6, 2019. phys.org/news/2019-06-planck-evidence-cosmic-anomalies.html

14. Lawrence M. Krauss, "The Energy of Empty Space that Isn't Zero" [Interview transcript], *Edge*, July 5, 2006. www.edge.org/conversation/the-energy-of-empty-space-that-isn-39t-zero

15. Note that anisotropy in the CMB that may indicate a universe that does not obey the Copernican principle can explain accelerated expansion without dark energy existing. The acceleration in this case occurs because of the different density of mass in the universe here versus far away. See introduction section in Daniel Friedmann, "An Accelerating Universe with No Dark Energy," *Open Access Library Journal*, 5.8 (August 2018). www.scirp.org/journal/PaperInforCitation.aspx?PaperID=86754

16. Rabbi Aryeh Kaplan, *Sefer Yetzirah—The Book of Creation: In Theory and Practice* (San Francisco: Weiser Books, 1997), 1:1 and p. 20.

17. (i) Aryeh Kaplan, *Sefer Yetzirah—The Book of Creation: In Theory and Practice* (San Francisco: Weiser Books, 1997), 1:5 p. 44.
    (ii) This text states that space is infinite and gives the impression that space is cubic in shape. However, the text is describing spiritual

strata. Physical space devolves from the creation of the void: a small point evenly spaced on all sides, not infinite, and giving the impression of a spherical shape. (As explained elsewhere, this is all in the spiritual strata and may or may not correspond to the physical world.)

[18] Rabbi Menachem Mendel of Lubavitch, *Derech Mitzvosecha—A Mystical Perspective on the Commandments* (Brooklyn, NY: Sichos in English, 2007), p. 350; available at http://www.hebrewbooks.org/15419 (English) and http://hebrewbooks.org/16082 (Hebrew).

[19] Rabbi Shmuel Schneersohn of Lubavitch, *True Existence* (Brooklyn, NY: Kehot Publication Society, 2002), p. 50.

[20] (i) Rabbi Shmuel Schneersohn of Lubavitch, *True Existence* (Brooklyn, NY: Kehot Publication Society, 2002), pp. 48–49, footnotes 85–86.
(ii) Some sources say this is the light of the luminaries, i.e., physical light (Chagigah 12a). Rashi says it is special and reserved for the End of Days—which implies it is Godly light. Both interpretations are correct at their particular level of understanding. The spiritual and physical lights appear in the same sentence and thus are counterparts: "the light of the first day was special. ... nonetheless the emanation of its potentials provided the illumination that was embodied in the luminaries of the 4th day. ... the luminaries which our God has created are good, He formed them with knowledge, discernment and wisdom. He endowed them with strength and power, that they may be a 'moshel' [translated as rule but also means analogy within the world]." If they are an analogy, the analogy is to God, so their light is analogous to the Ohr Ein Sof. From Meir Zlotowitz, Bereishis, *Genesis: A New Translation with a Commentary Anthologized from Talmudic Midrashic and Rabbinic Sources* (New York: Mesorah Publications Ltd., 1977), pp. 39–40.
(iii) Kli Yakar, Shlomo Ephraim ben Aaron Luntschitz: "the light of the first day was indeed special—per Rashi. Nevertheless, the emanations of its potentials provided the illumination that was embodied in the luminaries of the fourth day."

[21] Yitzchak Ginsburgh, *Lectures on Torah and Modern Physics* (Jerusalem: Gal Einai Publications, 2013), pp. 42–43. The analogy in Kabbalah between light and God says that "light clings to its source," meaning it is both the source and the destination, so it does not experience time. This is special relativity.

CHAPTER 11

1. See Daniel Friedmann and Dania Sheldon, *The Biblical Clock: The Untold Secrets Linking the Universe and Humanity with God's Plan* (USA: Inspired Books, 2019), Chapter 9.

2. Jacob Immanuel Schochet, *Mystical Concepts in Chassidism: An Introduction to Kabbalistic Concepts and Doctrines* (Brooklyn, NY: Kehot Publication Society, 1979), p. 143. www.scribd.com/doc/13658637/Schochet-Jacob-Immanuel-Mystical-Concepts-in-Chassidism-An-Introduction-to-Kabbalistic-Concepts-and-Doctrines

3. Yitzchak Ginsburgh, *Lectures on Torah and Modern Physics* (Jerusalem: Gal Einai Publications, 2013), p. 69. "General relativity is the world of tikun while quantum mechanics is an example of the world of tohu."

4. Unchanging strictly only from 5,781 years ago, as creations could evolve during the six days—equivalent to 13.8 billion years—and were only strictly fixed at the end of the six days.

5. "All composite particles in nature decay into proton[s] (which [are] made of quarks that are not found alone, electron[s], and electron type neutrino[s] which are the stable particles of nature"; Suresh Emre, "Why Elementary Particles Decay," *Renaissance Universal*, sureshemre.wordpress.com/2013/11/17/why-elementary-particles-decay

6. (i) Hirsch, Genesis 2.1, translates vayehulu as "they were brought to their conclusion."
   (ii) Based on Psalm 148:6: "And He set them up to eternity, yea forever, He issued a decree, which will not change." Rambam explains (*Guide for the Perplexed*, translated by M. Friedländer, 2nd edition (London: Routledge & Kegan Paul, 1904), Part 2, Chapter 30) that this rule became operative after the completion of the six days of creation. Before that time, the laws governing the universe could have been quite different from those we are familiar with today. From Rabbi Moshe Meiselman, *Torah, Chazal and Science* (Israel: Israel Bookshop Publications, 2013), p. 409.

7. Benjamin Schumacher, *Quantum Mechanics: The Physics of the Microscopic World Course Guidebook* (Chantilly, VA: The Teaching Company, 2009), Chapters 6 and 7.

8   Aryeh Kaplan, *Sefer Yetzirah—The Book of Creation: In Theory and Practice* (San Francisco: Weiser Books, 1997), pp. 145–146.

9   This is a speculative topic, but a few peer-reviewed articles have been published. For an overview, see https://www.livescience.com/55283-has-a-fifth-force-been-discovered.html

10   Yitzchak Ginsburgh, *The Hebrew Letters: Channels of Creative Consciousness* (Jerusalem: Gal Einai Publications, 1990).

11   Yitzchak Ginsburgh, *The Hebrew Letters: Channels of Creative Consciousness* (Jerusalem: Gal Einai Publications, 1990), p. 153.

12   Yitzchak Ginsburgh, *The Hebrew Letters: Channels of Creative Consciousness* (Jerusalem: Gal Einai Publications, 1990), p. 79.

13   Yitzchak Ginsburgh, *The Hebrew Letters: Channels of Creative Consciousness* (Jerusalem: Gal Einai Publications, 1990), p. 93.

CHAPTER 12

1   Daniel E. Friedmann, "A Complete Set of 22 Elementary Particles for an Expanded Standard Model (Version 2)," *Open Access Library Journal*, 7 (September 2020): 1–10. 10.4236/oalib.1106715

2   Aryeh Kaplan, *Sefer Yetzirah—The Book of Creation: In Theory and Practice* (San Francisco: Weiser Books, 1997), p. 29.

3   The actual correspondence of the diagram to the human body is different than explained herein, where I have chosen the easiest way to narrate the diagram.

4   Aryeh Kaplan, *Sefer Yetzirah—The Book of Creation: In Theory and Practice* (San Francisco: Weiser Books, 1997), pp. 30–33.

5   Aryeh Kaplan, *Sefer Yetzirah—The Book of Creation: In Theory and Practice* (San Francisco: Weiser Books, 1997), pp. 145–146.

6   Ethan Siegel, "Quarks Don't Actually Have Colors," *Forbes*, April 18, 2019. www.forbes.com/sites/startswithabang/2019/04/18/quarks-dont-actually-have-colors/#334386e31fe5

7   Aryeh Kaplan, *Sefer Yetzirah—The Book of Creation: In Theory and Practice* (San Francisco: Weiser Books, 1997), p. 200. Note that four letters produce twenty-four permutations, but two of the letters are the same, so in fact only twelve arrangements are produced.

8   Because these are diagonal, they interact with the three horizontal forces just discussed and the vertical force, gravity, thus they are visible. In contrast, the vertical letters, the doubles, only interact with the vertical force of gravity and are thus invisible or dark matter.

9   Thus, since these letters have two sounds, some or all of the seven particles could have doubles, i.e., sister particles of the same mass but with some other different characteristic(s). There is disagreement as to whether all the letters have two sounds, despite their name. Thus, it's not clear whether all particles have sister particles.

10 The sefirot come in three levels. The top level corresponds to the intellect, the second level to the primary emotions (e.g., kindness), and the third to the emotions we experience after an act (e.g., how we feel after giving charity). When brought down to the physical level, only the lowest level can manifest permanent things, which exist forever (in scientific terms, things that have infinite life or do not decay). The other levels are there only to give rise to the lowest level and are not permanent; these particles decay. The exception is particles of no mass that go at the speed of light and thus do not experience any time. To us, these also seem to exist forever; the photon is an example.

11 See, for example, Kyriakos Vattis, Savvas M. Koushiappas, and Abraham Loeb, "Dark Matter Decaying in the Late Universe Can Relieve the H0 Tension," *Physical Review D,* 99, 121302(R) (June 10, 2019), https://arxiv.org/abs/1903.06220

12 "The Matter-Antimatter Asymmetry Problem," CERN, home.cern/science/physics/matter-antimatter-asymmetry-problem

13 Rashi on Genesis 1:1.

14 Andro Barnaveli and Merab Gogberashvili, "Baryon and Time Asymmetries of the Universe," *Physics Letters*, 3168 (1993): 1–46. https://arxiv.org/abs/hep-ph/9505413

15 "Time's Arrow: Particles Cannot Go Back to the Future," CERN, November 6, 1998. home.cern/news/press-release/cern/times-arrow-particles-cannot-go-back-future

CHAPTER 13

1. For an understandable and more complete list, see en.wikipedia.org/wiki/List_of_unsolved_problems_in_physics

2. Some have gone back to the proposal that the universe is eternal or at least has gone and will be going through many cycles, this cycle being just one. In this scenario, inflation does not happen. The prior universe shrinks but remains large enough to be classical and re-emerge, avoiding the need for inflation. This scenario agrees with the Torah in the sense that this universe starts from a large enough bit of space and there is something eternal (in this case, the "universe"; in Torah's case, God). However, this proposal pushes the problem of the beginning to an earlier universe and time. See, for example, Anna Ijjas and Paul Steinhardt, "A New Kind of Cyclic Universe," *Physics Letters B*, 795 (2019). doi:10.1016/j.physletb.2019.06.056

3. M. Agostini et al., "Test of Electric Charge Conservation with Borexino," *Physical Review Letters* 115, 231802 (December 3, 2015). journals.aps.org/prl/abstract/10.1103/PhysRevLett.115.231802

4. Aryeh Kaplan, *Sefer Yetzirah—The Book of Creation: In Theory and Practice* (San Francisco: Weiser Books, 1997), 1:1 and p. 20.

CHAPTER 14

1. Alternatively, the four fundamental forces—electromagnetism, gravitation, weak nuclear interaction, and strong nuclear interaction—were one fundamental force. This force later split sequentially into the four forces, gravity being the first to appear from it.

2. (i) Anna Ijjas and Paul Steinhardt, "A New Kind of Cyclic Universe," *Physics Letters B*, 795 (2019). doi:10.1016/j.physletb.2019.06.056
   (ii) Joshua Howgego, "10 Mysteries of the Universe: How Did It All Begin?" *New Scientist* (September 19, 2018). www.newscientist.com/article/mg23931960-200-10-mysteries-of-the-universe-how-did-it-all-begin/#ixzz5z5o4Hasm
   (iii) Anna Ijjas, "What If There Was No Big Bang and We Live in an Ever-Cycling Universe?" *New Scientist*, August 14, 2019. www.newscientist.com/article/mg24332430-800-what-if-there-was-no-big-bang-and-we-live-in-an-ever-cycling-universe/#ixzz5z5ofVWST

3   Sean Carroll, "A Universe from Nothing?" *Discover*, April 28, 2012. http://blogs.discovermagazine.com/cosmicvariance/2012/04/28/a-universe-from-nothing/#.W6GDEaZKhPY

4   Exodus 3:14.

5   Rambam, fourth principle of faith. See https://www.chabad.org/library/article_cdo/aid/332555/jewish/Maimonides-13-Principles-of-Faith.htm

6   Rabbi Schneur Zalman of Liadi, *Iggeret Hakaodesh (the Tanya)*, bilingual edition (Brooklyn, NY: Kehot Publication Society, 1984), volume 4, epistle 20.

7   Midrash Rabbah Genesis 39.14.

8   Yitzchak Ginsburgh, *Lectures on Torah and Modern Physics* (Jerusalem: Gal Einai Publications, 2013), pp. 143–144.

9   The steps described in the middle column of the table are in the spiritual strata, not the physical. Our physical world is the culmination of contractions from the spiritual; Rabbi Menachem Mendel of Lubavitch, *Derech Mitzvosecha—A Mystical Perspective on the Commandments* (Brooklyn, NY: Sichos in English, 2007), p. 185; available at http://www.hebrewbooks.org/15419 (English) and http://hebrewbooks.org/16082 (Hebrew).
    The "physical is a reflection of the spiritual template"; *The Kabbalah of Time* (Brooklyn, NY: The Rohr Jewish Learning Institute, n.d.), lesson 2. Thus, it is possible that the creation of the physical world reflected that deeper spiritual origin; we make this assumption when generating the right-hand column in the table.

10 Rabbi Menachem Mendel of Lubavitch, *Derech Mitzvosecha—A Mystical Perspective on the Commandments* (Brooklyn, NY: Sichos in English, 2007), Chapters 11 and 12; available at http://www.hebrewbooks.org/15419 (English) and http://hebrewbooks.org/16082 (Hebrew).

11 The size of universe is a few millimeters after the breaking of the vessels, thus macro and micro never coexist.

12 Rabbi Schneur Zalman of Liadi, *Likutei Amarim, Part Two: Shaar Hayichud Vehaemuna*, bilingual edition (Brooklyn, NY: Kehot Publication Society, 1993), Chapter 2; available at http://www.chabad.org/library/tanya/tanya_cdo/aid/1029162/jewish/Shaar-Hayichud-Vehaemuna.htm

13 Rabbi Schneur Zalman of Liadi, *Iggeret Hakaodesh (the Tanya)*, bilingual edition (Brooklyn, NY: Kehot Publication Society, 1984), volume 4, epistle 20.

14 Rabbi Schneur Zalman of Liadi, *Iggeret Hakaodesh (the Tanya)*, bilingual edition (Brooklyn, NY: Kehot Publication Society, 1984), volume 4, epistle 20.

15 Rabbi Schneur Zalman of Liadi, *Iggeret Hakaodesh (the Tanya)*, bilingual edition (Brooklyn, NY: Kehot Publication Society, 1984), volume 4, epistle 20.

CHAPTER 15

1 Anil Ananthaswamy, "Atomic Clocks Make Best Measurement Yet of Relativity of Time," *New Scientist* (March 22, 2017). www.newscientist.com/article/mg23331184-900-atomic-clocks-make-best-measurement-yet-of-relativity-of-time/

2 That it's homogeneous and the same density everywhere at any given time.

3 Carlo Rovelli, *The Order of Time* (New York: Penguin, 2018), Chapter 13.

4 Lubavitcher Rebbe, adapted by Yanki Tauber, *Inside Time: A Chassidic Perspective on the Jewish Calendar* (Brooklyn, NY: Meaningful Life Center, 2015), volume 1, p. 213.

5 Past, present, and future all exist simultaneously in the ultimate divine reality, expressed by God's essential four-letter name (*Havayah*), an amalgam of the words *hayah*, *hoveh*, and *yihyeh*—"was, is, and will be." See Rabbi Menachem Mendel of Lubavitch, *Derech Mitzvosecha—A Mystical Perspective on the Commandments* (Brooklyn, NY: Sichos in English, 2007), Chapter 12; available at http://www.hebrewbooks.org/15419 (English) and http://hebrewbooks.org/16082 (Hebrew).

6 Lubavitcher Rebbe, adapted by Yanki Tauber, *Inside Time—A Chassidic Perspective on the Jewish Calendar* (Brooklyn, NY: Meaningful Life Center, 2015), volume 1, p. 10.

7 Moses Maimonides, *Guide for the Perplexed*, translated by M. Friedländer, 2nd edition (London: Routledge & Kegan Paul, 1904), Book 2, Chapter 13. http://www.sacred-texts.com/jud/gfp/gfp100.htm

8   Ramban (Nachmanides), translated and annotated by Charles B. Chavel, *Commentary on the Torah, Genesis* (New York: Shilo Publishing House, 1971), p. 25.

9   Rabbi Menachem Mendel of Lubavitch, *Derech Mitzvosecha—A Mystical Perspective on the Commandments* (Brooklyn, NY: Sichos in English, 2007), Chapters 11 and 12; available at http://www.hebrewbooks.org/15419 (English) and http://hebrewbooks.org/16082 (Hebrew).

10 For an explanation, see Daniel Friedmann, *Roadmap to the End of Days*, introduction and Chapter 3.

11 The hour has a special meaning in Jewish law. For example, "the third hour of the day" doesn't mean 3am. Rather, an hour is calculated by taking the total time of daylight of a particular day, from sunrise until sunset, and dividing it into twelve equal parts. This "proportional hour" thus varies by season, by location, and even by day. Similarly, nighttime is divided into twelve periods or proportional hours.

12 The essential name of God has three unique letters (two of its four letters are the same), which can be arranged in twelve different orders. Similarly, the other name also has three unique letters, again yielding twelve permutations.

13 Rabbi Menachem Mendel of Lubavitch, *Derech Mitzvosecha—A Mystical Perspective on the Commandments* (Brooklyn, NY: Sichos in English, 2007), Chapter 12; available at http://www.hebrewbooks.org/15419 (English) and http://hebrewbooks.org/16082 (Hebrew).

14 Rabbi Menachem Mendel of Lubavitch, *Derech Mitzvosecha—A Mystical Perspective on the Commandments* (Brooklyn, NY: Sichos in English, 2007), Chapters 11 and 12; available at http://www.hebrewbooks.org/15419 (English) and http://hebrewbooks.org/16082 (Hebrew).

15 Genesis 1–14.

16 Lubavitcher Rebbe, adapted by Yanki Tauber, *Inside Time—A Chassidic Perspective on the Jewish Calendar* (Brooklyn, NY: Meaningful Life Center, 2015), volume 1, p. 10.

17 Lubavitcher Rebbe adapted by Yanki Tauber, *Inside Time—A Chassidic Perspective on the Jewish Calendar* (Brooklyn, NY: Meaningful Life Center, 2015), volume 1, appointments in time, pp. 38–45.

18 See Daniel Friedmann, *Roadmap to the End of Days*, for a full explanation of this concept applied to world history.

Endnotes

19 *The Kabbalah of Time* (Brooklyn, NY: The Rohr Jewish Learning Institute, n.d.), p. 57.

20 See Daniel Friedmann, *Roadmap to the End of Days*, Appendix B, for an explanation of the correspondence of Creation Days and millennia and the characteristics of time.

21 For a full treatment, see Lubavitcher Rebbe adapted by Yanki Tauber, *Inside Time—A Chassidic Perspective on the Jewish Calendar* (Brooklyn, NY: Meaningful Life Center, 2015).

22 Aryeh Kaplan, *Sefer Yetzirah—The Book of Creation: In Theory and Practice* (San Francisco: Weiser Books, 1997), 1:5, p. 49.

23 Rabbi Moshe Shilat, "World, Year, Soul," 1475, Chapter Behaalotcha, 16th of Sivan 5773 (May 25, 2013). http://www.zomet.org.il/eng/?CategoryID=160

24 (i) Some sources say this is the light of the luminaries, i.e., physical light (Chagigah 12a). Rashi says it is special and reserved for the End of Days—which implies it is Godly light. Both interpretations are correct at their particular level of understanding. The spiritual and physical lights appear in the same sentence and thus are counterparts: "the light of the first day was special. ... nonetheless the emanation of its potentials provided the illumination that was embodied in the luminaries of the 4th day. ... the luminaries which our God has created are good, he formed them with knowledge, discernment and wisdom. He endowed them with strength and power, that they may be a 'moshel' [translated as rule but also means analogy within the world]." If they are an analogy, the analogy is to God, so their light is analogous to the Ohr Ein Sof. From Meir Zlotowitz, Bereishis, *Genesis: A New Translation with a Commentary Anthologized from Talmudic Midrashic and Rabbinic Sources* (New York: Mesorah Publications Ltd., 1977), pp. 39–40.

   (ii) Kli Yakar, Shlomo Ephraim ben Aaron Luntschitz: "the light of the first day was indeed special—per Rashi. Nevertheless, the emanations of its potentials provided the illumination that was embodied in the luminaries of the fourth day."

25 Yitzchak Ginsburgh, *Lectures on Torah and Modern Physics* (Jerusalem: Gal Einai Publications, 2013), pp. 42–43: the analogy in Kabbalah between light and God says that "light clings to its source," meaning it is both the source and the destination. This is special relativity, whereby light does not experience time.

CHAPTER 16

1. Benjamin Schumacher, *Quantum Mechanics: The Physics of the Microscopic World*. Course Guidebook (Chantilly, VA: The Teaching Company, 2009), Chapters 6 and 7.

2. https://www.newscientist.com/article/mg23130820-200-collapse-has-quantum-theorys-greatest-mystery-been-solved/

3. https://www.newscientist.com/article/mg23130820-200-collapse-has-quantum-theorys-greatest-mystery-been-solved/

4. Rabbi Schneur Zalman of Liadi, *Likutei Amarim, Part Two: Shaar Hayichud Vehaemuna*, bilingual edition (Brooklyn, NY: Kehot Publication Society, 1993), Chapter 7; available at www.chabad.org/library/tanya/tanya_cdo/aid/1029162/jewish/Shaar-Hayichud-Vehaemuna.htm

5. Tohu starts with Genesis 1:2 and tikkun starts with Genesis 1:3. Jacob Immanuel Schochet, *Mystical Concepts in Chassidism: An Introduction to Kabbalistic Concepts and Doctrines* (Brooklyn, NY: Kehot Publication Society, 1979), p. 143.
https://www.scribd.com/doc/13658637/Schochet-Jacob-Immanuel-Mystical-Concepts-in-Chassidism-An-Introduction-to-Kabbalistic-Concepts-and-Doctrines

6. Yitzchak Ginsburgh, *Lectures on Torah and Modern Physics* (Jerusalem: Gal Einai Publications, 2013), p. 69. "General relativity is the world of tikun while quantum mechanics is an example of the world of tohu."

7. Jacob Immanuel Schochet, *Mystical Concepts in Chassidism: An Introduction to Kabbalistic Concepts and Doctrines* (Brooklyn, NY: Kehot Publication Society, 1979), Chapter 7.
https://www.scribd.com/doc/13658637/Schochet-Jacob-Immanuel-Mystical-Concepts-in-Chassidism-An-Introduction-to-Kabbalistic-Concepts-and-Doctrines

8. (i) Jacob Immanuel Schochet, *Mystical Concepts in Chassidism: An Introduction to Kabbalistic Concepts and Doctrines* (Brooklyn, NY: Kehot Publication Society, 1979), Chapters 7–9.
https://www.scribd.com/doc/13658637/Schochet-Jacob-Immanuel-Mystical-Concepts-in-Chassidism-An-Introduction-to-Kabbalistic-Concepts-and-Doctrines
(ii) Gershom Scholem, *Kabbalah* (New York: Times Books, 1978), pp. 135–144.

9   Yitzchak Ginsburgh, *Lectures on Torah and Modern Physics* (Jerusalem: Gal Einai Publications, 2013), p. 69.

10 Yitzchak Ginsburgh, *Lectures on Torah and Modern Physics* (Jerusalem: Gal Einai Publications, 2013), pp. 70–71.

11 Benjamin Schumacher, *Quantum Mechanics: The Physics of the Microscopic World.* Course Guidebook (Chantilly, VA: The Teaching Company, 2009), Chapter 7.

12 Benjamin Schumacher, *Quantum Mechanics: The Physics of the Microscopic World.* Course Guidebook (Chantilly, VA: The Teaching Company, 2009), Chapters 6 and 7.

13 Yitzchak Ginsburgh, *137 The Riddle of Creation* (Jerusalem: Gal Einai, 2019), pp. 153–154.

14 Richard Wolfson, *Einstein's Relativity and the Quantum Revolution* (Chantilly, VA: The Teaching Company, 2000), Chapter 24.

CHAPTER 17

1   Brian Greene, *The Fabric of the Cosmos: Space, Time and the Texture of Reality* (New York: Vintage Books, 2005), p. 353.

2   "The Matter-Antimatter Asymmetry Problem," CERN, home.cern/science/physics/matter-antimatter-asymmetry-problem

3   Psalm 33:9.

4   Rabbi Schneur Zalman of Liadi, *Likutei Amarim, Part Two: Shaar Hayichud Vehaemuna*, bilingual edition (Brooklyn, NY: Kehot Publication Society, 1993), Chapter 7; available at http://www.chabad.org/library/tanya/tanya_cdo/aid/1029162/jewish/Shaar-Hayichud-Vehaemuna.htm

5   For example, the letters can be transposed so that the first letter is interchanged with the last letter of the alphabet and the second with the second to last, etc. to generate an equivalent transposed word. This transposed word relates to the original word but brings something else into being.

6   In Figure 7.4, we saw the normative values of the letters, but there are other systems. For example, the ordinal value is when each letter is given a value of 1 to 22 in order; the reduced value is obtained from the normative value by adding the numerals. Thus, 10 becomes 1 and 20 becomes 2 and so on, with many other more complex systems.

7. Rabbi Schneur Zalman of Liadi, *Likutei Amarim, Part Two: Shaar Hayichud Vehaemuna*, bilingual edition (Brooklyn, NY: Kehot Publication Society, 1993), Chapter 7, p. 937; available at http://www.chabad.org/library/tanya/tanya_cdo/aid/1029162/jewish/Shaar-Hayichud-Vehaemuna.htm

8. Yitzchak Ginsburgh, *Kabbalah and Meditation for the Nations* (Jerusalem: Gal Einai Publications, 2007), p. 69.

9. For example: The electron, with its well-defined mass, is used as the control factor to compute the matter particle masses. The electron corresponds to the letter yod, whose numerical value is 10. The scale factor for level 1 (lowest energy) on the right-side pillar is the electron mass divided by 10 or 0.0511. To obtain the scale factor for level 1 on the left-side pillar we multiply by the pillar ratio 365/248 and obtain 0.0752. Then, to obtain the mass of the down quark corresponding to the letter samech we multiply the scale factor of 0.0752 by the letter's numerical value of 60, and obtain a mass of 4.51 MeV/c^2.

   For the derivation of all particle masses see Daniel E. Friedmann, "A Complete Set of 22 Elementary Particles for an Expanded Standard Model (Version 2)," *Open Access Library Journal*, 7 (September 2020): 1–10. 10.4236/oalib.1106715

10. P.A. Zyla *et al.* (Particle Data Group), to be published in Prog. Theor. Exp. Phys. 2020, 083C01 (2020). https://pdg.lbl.gov/2020/tables/contents_tables.html

11. Note a method for predicting the masses of the three almost massless leptons has not been developed at this time.

12. Daniel E. Friedmann, "A Complete Set of 22 Elementary Particles for an Expanded Standard Model (Version 2)," *Open Access Library Journal*, 7 (September 2020): 1–10. 10.4236/oalib.1106715

CHAPTER 18

1. Richard Feynman, *The Character of Physical Law* (Cambridge, MA: MIT Press, 1985), pp. 54–55. dillydust.com/The%20Character%20of%20Physical%20Law~tqw~_darksiderg.pdf

2. Rabbi Sholom Dovber of Lubavitch, Chassidic discourse Heichaltzu 5659 (1899).

3   Daniel Friedmann, *The Broken Gift* (USA: Inspired Books, 2013), Chapters 6 and 8.

4   (i) For a scientific description, see Sally McBrearty and Alison S. Brooks, "The Revolution that Wasn't: A New Interpretation of the Origin of Modern Human Behavior," *Journal of Human Evolution*, 39 (2000): 453–563.
    (ii) For the Biblical description of what comes with the Divine soul, see Daniel Friedmann, *The Broken Gift* (USA: Inspired Books, 2013), Chapter 6.

5   Adapted from: (i) Rabbi Schneur Zalman of Liadi, "Introducing The Gate to Understanding G-d's Unity and the Faith," nathanielsegal.mysite.com/Tanya2.html#TEXT; (ii) Rabbi Menachem Mendel of Lubavitch, *Derech Mitzvosecha—A Mystical Perspective on the Commandments* (Brooklyn, NY: Sichos in English, 2007), Chapter 6; available at http://www.hebrewbooks.org/15419 (English) and http://hebrewbooks.org/16082 (Hebrew).

6   Deuteronomy 6:4.

7   Maimonides, Yad HaChazakah, "Hilchoth Yesodei Hatorah," 1:7. Cf. Moreh Nevuchim, I, Chapters 50 and 53.

8   Rabbi Schneur Zalman of Liadi, *Torah Or* (Brooklyn, NY: Kehot Publication Society, 1954) [Hebrew], p. 166a. This rules out all pantheistic doctrines that would identify or compound God with nature.

9   Although everything is divinity, this does not mean God is identical with the world or limited by it. He is immanent in Creation, yet at the same time He infinitely transcends it. All existence is brought into being and sustained through the divine speech, the words and letters of the Ten Utterances (the ten sayings of Creation, such as "Let there be light," found in Genesis, Chapter 1) by which the world was created. Existence is as nothing in relation to His Infinite Being. The divine speech is united with His essence in a complete union even after it has become "materialized" in the creation of the world. Thus, created beings are always "within" their source; they appear as independently existing entities only from the perspective of created beings who are incapable of perceiving their source.

10 Isaiah 6:3.

APPENDIX A

1 Zyla, P.A., et al. (Particle Data Group) (2020) Review of Particle Physics. Progress of Theoretical and Experimental Physics, 2020, 083C01. https://pdg.lbl.gov/2020/html/authors_2020.html

APPENDIX B

1 Updated paraphrase of Jacob Immanuel Schochet, *Mystical Concepts in Chassidism: An Introduction to Kabbalistic Concepts and Doctrines* (Brooklyn, NY: Kehot Publication Society, 1979), pp. 41–45; available at www.hebrewbooks.org/15600

www.ingramcontent.com/pod-product-compliance
Lightning Source LLC
Chambersburg PA
CBHW060824220526
45466CB00003B/967